Algebra and Tiling

Homomorphisms in the Service of Geometry

The Carus Mathematical Monographs

Number Twenty-five

Algebra and Tiling

Homomorphisms in the Service of Geometry

Sherman K. Stein
University of California, Davis
and
Sándor Szabó
University of Bahrain

Published and Distributed by
THE MATHEMATICAL ASSOCIATION OF AMERICA

© *1994 by*
The Mathematical Association of America (Incorporated)
Library of Congress Catalog Card Number 94-79589

Complete Set ISBN 0-88385-000-1
Vol. 25 ISBN 0-88385-028-1

Printed in the United States of America

Current Printing (last digit):
10   9   8   7   6   5   4   3   2   1

THE
CARUS MATHEMATICAL MONOGRAPHS

Published by
THE MATHEMATICAL ASSOCIATION OF AMERICA

The following Monographs have been published:

Mathematical Association of America
1529 Eighteenth Street, NW
Washington, DC 20036
800-331-1MAA   FAX: 202-265-2384

Preface

If n-dimensional space is tiled by a lattice of parallel unit cubes, must some pair of them share a complete $(n-1)$-dimensional face?

Is it possible to tile a square with an odd number of triangles, all of which have the same area?

Is it possible to tile a square with 30°-60°-90° triangles?

For positive integers k and n a (k, n)-semicross consists of $kn + 1$ parallel n-dimensional unit cubes arranged as a corner cube with n arms of length k glued on to n non-opposite faces of the corner cube. (If n is 2, it resembles the letter L, and, if n is 3, a tripod.) For which values of k and n does the (k, n)-semicross tile space by translates?

The resolution of each of these questions quickly takes us away from geometry and places us in the world of algebra.

The first one, which grew out of Minkowski's work on diophantine approximation, ends up as a question about finite abelian groups, which is settled with the aid of the group ring, characters of abelian groups, factor groups, and cyclotomic fields.

Tiling by triangles of equal areas leads us to call on valuation theory and Sperner's lemma, while tiling by similar triangles turns out to involve isomorphisms of subfields of the complex numbers.

The semicross forces us to look at homomorphisms, cosets, factor groups, number theory, and combinatorics.

Of course there is a long tradition of geometric questions requiring algebra for their answers. The oldest go back to the Greeks:

"Can we trisect every angle with straightedge and compass?" "Can we construct a cube with twice the volume of a given cube?" "Can we construct a square with the same area as that of a given disk?" These were not resolved until we had the notion of the dimension of a field extension and also knew that π is transcendental.

We consider only the algebra that has been used to solve tiling and related problems. Even so, we do not cover all such problems. For instance, we do not describe Conway's application of finitely presented groups to tiling by copies of a given figure. See [19] in the Bibliography on pp. 200– 201. Nor do we treat Barnes' use of algebraic geometry [1]. Thurston [22] has written a nice exposition of Conway's work, and providing the algebraic background for Barnes' work would take too many pages. The group with generators a and b and relations $a^2 = e = b^3$ plays a key role in obtaining the Banach–Tarski paradox, which asserts that a pea can be divided into a finite number of pieces that can be reassembled to form the sun. A clear exposition of the argument was given by Meschkowski [16].

We had two types of readers in mind as we wrote, the undergraduate or graduate student who has had at least a semester of algebra, and the experienced mathematician. To make the exposition accessible to the beginner we have added a few appendices that cover some special topics not usually found in a typical introductory algebra course and also included exercises to serve as a study guide. For both the beginner and the expert we include questions that have not yet been answered, which we call "Problems," to distinguish them from the exercises.

Now a word about the organization of this book and the order in which the chapters may be read.

Chapter 1 describes the history leading up to Minkowski's conjecture on tiling by cubes. We give the solution of that problem in the form of Rédei's broad generalization of Hajós's original solution. Its proof, which is much longer than the proofs in the other chapters, is delayed until Chapter 7. (However, the proof uses only such basic notions as finite abelian groups, factor groups, homomorphisms from abelian groups into the complex numbers, finite fields, and polynomials over those fields.)

The beginner might start with Chapter 1, go to de Bruijn's harmonic bricks in Chapter 2, and then move on to Chapters 3 and 4, which concern the semicross and its centrally symmetric companion, the cross. After that, Chapters 5 and 6, which concern tiling by triangles, can be taken in either order. With the experience of studying these chapters, the beginner would then be ready for the rest of Chapter 2 and the proof of Rédei's theorem. The advanced reader may examine the chapters in any order, since they are essentially independent.

We hope that instructors will draw on these chapters in their algebra courses, in order to bring abstract algebra ideas down to earth by applying them in geometric settings.

The little that we include in the seven chapters is only the tip of the iceberg. The references at the end of each chapter and the bibliography at the end of the book will enable the reader to pursue the topics much further.

Several of these describe quite recent work. In [14] Laczkovich and Szekeres obtain the following result. Let r be a positive real number. Then a square can be tiled by rectangles whose width and length are in the ratio r if and only if r is algebraic and the real parts of its conjugates are positive. This was done in 1990. Independently, Freiling and Rinne obtained the same result in 1994 by similar means. Kenyon [12] considers the question: Which polygons can be tiled by a finite number of squares? Gale [8] obtains a short proof using matrices of Dehn's theorem, which asserts that in a tiling of a unit square by a finite number of squares all the squares have rational sides.

We wish to acknowledge the contributions of several people to this book. Victor Klee, Miklós Laczkovich, Lajos Pósa, Fred Richman, and John Thomas, in response to our requests, described the background of their work. Mark Chrisman and Dean Hickerson graciously permitted us to include some of their unpublished proofs. Aaron Abrams, while a senior at the University of California at Davis, read two earlier versions of the manuscript and made many suggestions that should help the book serve a broader audience. The exposition also benefited from the advice of G. Donald Chakerian and Marjorie Senechal.

Budapest   Davis, California
August, 1994   August, 1994

Contents

Chapter 1

Minkowski's Conjecture

The origins of most of the tiling problems we will explore go back just a short time. But Minkowski's problem, like many ideas in mathematics, can trace its roots to the Pythagorean theorem, $a^2 + b^2 = c^2$. We shall trace these roots, and then go on to describe some of the problems that Minkowski's question, in turn, suggested. We do this in order to place the problem in its proper position in the intricate web that constitutes mathematics.

1. Introduction

The Greeks discovered all integer solutions of the equation $x^2 + y^2 = z^2$: pick three integers d, m, and n, and let one of x and y be $d(2mn)$ and the other be $d(m^2 - n^2)$, and $z = d(m^2 + n^2)$. Inspired by this, Diophantos, who lived some time between A.D. 150 and A.D. 350, described techniques for producing rational solutions of such equations as $x^2 + y^2 = z^3$, $x^3 + y^3 - z^2$, and $x^4 + y^4 = z^3$. He does not refer to the equation $x^3 + y^3 = z^3$, whose first recorded mention is in a letter written about the year A.D. 1000 by the mathematician Abu Djafar Mohammed Ben Alhocain [22]:

> I have already explained that the arguments that Abu Mohammed Alkhodjandi, may God have mercy on him, has proposed in this demonstration that on the addition

of two cube numbers there does not result a cube, are defective and inexact.

Alkhodjandi is better known for his observation of the obliqueness of the earth's orbit in 992.

Fermat, who was familiar with Diophantos through Bachet's edition of 1621, added a fresh twist, investigating the equation $x^2 + y^2 = p$, where x and y are integers and p is a prime. In a letter to Mersenne, in 1640, he wrote, "Each prime number that is one more than a multiple of four is uniquely the sum of two squares and uniquely the hypotenuse of a right triangle." Six months later, he confessed to Frenicle, "Finding those two squares is as difficult as groping in the dark."

But Fermat did not stop with the form $x^2 + y^2$. In a letter to Pascal in 1654 he stated, "Each prime that is one more than a multiple of 3 is composed of a square and the triple of another square, such as 7, 13, 19, 31, 37, etc.

Each prime that exceeds a multiple of 8 by 1 or 3 is composed of a square and twice another square, such as 11, 17, 19, 41, 43, etc. .... This will be followed by the invention of many propositions which do not appear in Diophantos."

In short, Fermat introduced the study of integers represented by the quadratic forms $x^2 + y^2$, $x^2 + 3y^2$, and $x^2 + 2y^2$. Euler went further, examining forms of the type $Ax^2 + Cy^2$, where A and C are fixed integers. Lagrange in 1775 studied the most general quadratic form $Ax^2 + Bxy + Cy^2$, where A, B, and C are fixed integers. If the integers A, B, and C have the property that $Ax^2 + Bxy + Cy^2$ is positive whenever x and y are not both 0, the form is called *positive definite*. Lagrange investigated the minimum nonzero value of such a form when x and y are integers. An estimate of this minimum would help in his classification of quadratic forms and, as we will show in the next section, it turns out that this minimum is related to the geometry of certain regularly spaced points in the plane, called "lattices."

2. Quadratic forms and lattices

Informally speaking, a lattice L in the plane, R^2, is a "homogeneous" set of isolated points. It is an example of a "regular" point set in that it looks the same no matter from which of its points you observe it. Figure 1 shows part of a lattice. (The whole lattice is unbounded.) We also assume that the origin is one of the lattice points.

Say that you are at a point $P \in L$ and a friend is at point $Q \in L$. You see a point $R \in L$. Then your friend at Q will be able to see a point at $Q + (R - P)$. (Here we treat points like vectors.) This is the algebraic translation of the homogeneity of the lattice, as shown in Figure 2.

Thus the lattice is closed under the operation "$Q + (R - P)$." Now let $Q = (0, 0)$. Then we see that the lattice is closed under the operation of subtraction, $R - P$. This tells us that L is a subgroup of the additive group of the vector space R^2. The other property we want of a lattice is that its points are not arbitrarily close to each other. This brings us to the formal definition of a lattice.

A set of points L in R^n is a *lattice* if it is a group under vector addition and each of its points is the center of a ball that contains no other points of L.

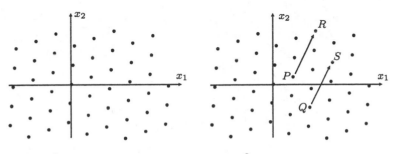

FIGURE 1

FIGURE 2

In Appendix A it is shown that if L is a lattice in n-space, then there are linearly independent vectors $v_1, v_2, \ldots, v_k, k \leq n$, such that L consists of all the points of the form

$$m_1 v_1 + m_2 v_2 + \cdots + m_k v_k,$$

where $m_i \in Z, 1 \leq i \leq k$. Such a set of vectors is called a *basis* for L, and k is called the *dimension* of the lattice. Figure 3 shows a basis for the lattice of Figure 1.

When you have a basis it is tempting to draw the dissection of the plane by the parallelograms that they suggest, as in Figure 4.

A lattice has an infinite number of bases (as Exercise 1 suggests). Figure 5 shows another basis for the lattice of Figure 1 and the associated parallelograms.

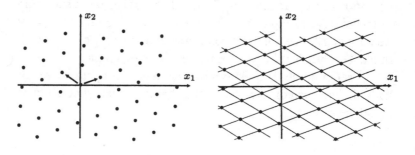

FIGURE 3   **FIGURE 4**

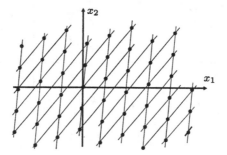

FIGURE 5

Let $v_1, v_2, \ldots, v_n$ be a basis for an n-dimensional lattice L in n-space. The parallelepiped whose edges are $v_1, v_2, \ldots, v_n$ is the set of points

$$\{x_1 v_1 + x_2 v_2 + \cdots + x_n v_n : 0 \le x_i \le 1, 1 \le i \le n\}.$$

It is called a *fundamental parallelepiped* of the lattice. The volume of this parallelepiped is the absolute value of the determinant of the matrix whose rows are the vectors $v_1, v_2, \ldots, v_n$. In Appendix A it is shown that this volume is independent of the particular basis chosen for L. It is called the *determinant* of L and is denoted d.

The most famous lattice in n-space is the set of points all of whose coordinates relative to the usual cartesian axes are integers, the group Z^n, called the *standard lattice*. The most conspicuous basis for this lattice consists of the n unit vectors

$$(1, 0, \ldots, 0), (0, 1, 0, \ldots, 0), \ldots, (0, \ldots, 1).$$

The determinant of Z^n is 1.

Exercise 1.
(a) Draw three different bases for the lattice Z^2 and their associated parallelograms.
(b) Check that in each case the determinant of the matrix whose rows are the basis is 1 or -1.

Exercise 2. Let a and b be integers. When is the vector (a, b) part of a basis for Z^2?

In a book review in 1831 Gauss [4] showed that positive definite quadratic forms in two variables are intimately connected with plane lattices. Gauss assumed that the coefficient of xy in the quadratic form is even. This is no loss of generality, since we could, after all, double all the coefficients of a quadratic form without losing any information about its minimum value. We first present his discovery as he described it. Then we will express it in modern terms.

The positive definite binary form $axx + 2bxy + cyy$ represents in general the square of the distance between arbitrary points in a plane whose coordinates along axes that meet at an angle whose cosine is $b/\sqrt{ac}$ are $x\sqrt{a}$ and $y\sqrt{c}$. In so far as x and y refer only to whole numbers, this form is a system of points that lie in the intersection of two systems of parallel lines. The lines of each system lie at equal distances from each other, and this common distance is $\sqrt{a}$, if measured parallel to the lines of the second system; the common distance of the other system, measured parallel to the lines of the first system, is $\sqrt{c}$. The angle between the two systems is the one specified above.

In this way the plane is divided into congruent parallelograms whose vertices form the system of points where none of the points can fall within the parallelograms. The determinant taken with a positive sign, that is, $ac - bb$, gives the square of the area of an elementary parallelogram.

One and the same system of such points can be divided into the vertices of a family of parallelograms in infinitely many different ways and thus referred back to equally many different forms, but all these forms are what is technically called "equivalent" and the area of those elementary parallelograms remains the same.

Figure 6 shows what Gauss had in mind.

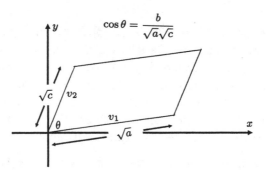

FIGURE 6

The square of the length of $xv_1 + yv_2$ is the value of the dot product

$$(xv_1 + yv_2) \cdot (xv_1 + yv_2),$$

which expands to

$$x^2(v_1 \cdot v_1) + 2xy(v_1 \cdot v_2) + y^2(v_2 \cdot v_2).$$

Since

$$v_1 \cdot v_1 = a, \qquad v_2 \cdot v_2 = c,$$

and

$$v_1 \cdot v_2 = |v_1||v_2| \cos\theta = \sqrt{a}\sqrt{c}(b/\sqrt{ac}),$$

the form $ax^2 + 2bxy + cy^2$ indeed does represent the square of the length of a typical vector in the lattice.

Exercise 3.
(a) Show that if $ax^2 + 2bxy + cy^2$ is positive definite, then $a > 0$, $c > 0$ and $b^2 < ac$.
(b) Is the converse true?

So, instead of looking for the minimum nonzero value of a positive definite quadratic form, we may examine the square of the length of the shortest nonzero vector in a lattice.

The form $ax^2 + 2bxy + cy^2$ in matrix notation reads

$$(x, y) \begin{pmatrix} a & b \\ b & c \end{pmatrix} \begin{pmatrix} x \\ y \end{pmatrix}.$$

More generally, a quadratic form in n variables,

$$\sum_{i=1}^{n} \sum_{j=1}^{n} a_{ij} x_i x_j \qquad \text{with} \quad a_{ij} = a_{ji},$$

can be written as

$$(x_1, x_2, \ldots, x_n) \begin{pmatrix} a_{11} & a_{12} & \cdots & a_{1n} \\ a_{21} & a_{22} & \cdots & a_{2n} \\ \vdots & \vdots & \ddots & \vdots \\ a_{n1} & a_{n2} & \cdots & a_{nn} \end{pmatrix} \begin{pmatrix} x_1 \\ x_2 \\ \vdots \\ x_n \end{pmatrix}.$$

Denoting the row vector $(x_1, x_2, \ldots, x_n)$ by x, its transpose (a column vector) by x^T, and the matrix (a_{ij}) by A, we see that a quadratic form is denoted by the matrix product xAx^T. The determinant of A is called the *determinant* of the quadratic form, which we denote D.

The matrix A is symmetric. Thus there is an orthogonal matrix O such that $OAO^T = E$, a diagonal matrix. Moreover, if xAx^T is positive definite, then the diagonal entries in E are positive. From this it follows that there is a nonsingular matrix B such that $A = BB^T$.

Exercise 4.
(a) Fill in the details of the previous argument.
(b) Prove that if B is a nonsingular n by n matrix, then the quadratic form xBB^Tx^T is positive definite.

The quadratic form $xAx^T = xBB^Tx^T$ can be rewritten as $(xB)(xB)^T$, which is simply the square of the length of the vector xB. If we restrict the coordinates of the vector x to be integers, then the set of vectors of the form xB forms a lattice. To be specific, let $B = (b_{ij})$. Then

$$xB = (x_1, x_2, \ldots, x_n) \begin{pmatrix} b_{11} & b_{12} & \cdots & b_{1n} \\ b_{21} & b_{22} & \cdots & b_{2n} \\ \vdots & \vdots & \ddots & \vdots \\ b_{n1} & b_{n2} & \cdots & b_{nn} \end{pmatrix}$$

$$= (x_1b_{11} + x_2b_{21} + \cdots + x_nb_{n1}, \ldots,$$

$$x_1b_{1n} + x_2b_{2n} + \cdots + x_nb_{nn})$$

$$= x_1(b_{11}, \ldots, b_{1n}) + \cdots + x_n(b_{n1}, \ldots, b_{nn}).$$

Denoting the jth row of B by v_j, we see that $xB = x_1v_1 + \cdots + x_nv_n$. Hence as x varies, xB sweeps out the lattice with basis $v_1, v_2, \ldots, v_n$.

Exercise 5.   What is the relation between the determinant D of a quadratic form, that is, the determinant of A, and the volume of the parallelepiped spanned by $v_1, v_2, \ldots, v_n$, that is, the absolute value of the determinant of B?

Exercise 6. Consider the lattice Z^2 that consists of the points in R^2 both of whose coordinates are integers.
(a) One basis for this lattice is $v_1 = (1, 0)$ and $v_2 = (0, 1)$. What is the associated quadratic form?
(b) Another basis is $v_1 = (2, 3)$ and $v_2 = (1, 1)$. What is the associated quadratic form in this case?

Exercise 7. Figure 7 indicates the plane tiled by equilateral triangles of side length 1.
(a) Give a simple basis for the set of vertices of the triangles.
(b) What is its determinant?
(c) What is the positive definite quadratic form associated with this lattice and the basis chosen in (a)?

Hermite in 1845 proved that a positive definite quadratic form in n variables with determinant 1 assumes a nonzero value less than or equal to $(4/3)^{(n-1)/2}$ for some integer vector. His proof was an

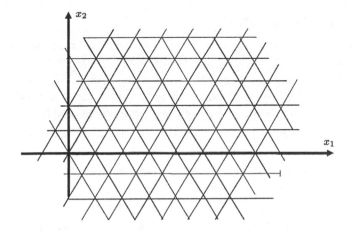

FIGURE 7

induction on n. For $n = 2$, this bound is $\sqrt{4/3} \approx 1.15$. In particular, the quadratic form $2x^2 + 10xy + 13y^2$ must assume the value 1 for some integers x and y.

Exercise 8. Find integers x and y such that $2x^2 + 10xy + 13y^2 = 1$.

Exercise 9. What does Hermite's theorem imply when the condition "determinant 1" is replaced by the condition "determinant D"?

Exercise 10. What does Hermite's theorem imply about a lattice of determinant 1 in the plane?

Exercise 11. What is the relation between the length of the shortest nonzero vector in a plane lattice and the largest radius of any set of congruent nonoverlapping disks whose centers are at the points of the lattice? (The problem of finding the length of the shortest vector in an n-dimensional lattice is thus intimately connected with the problem of finding how densely one may place nonoverlapping balls in n-space.)

3. Minkowski's theorem

Minkowski, introducing a new approach, improved on Hermite's bound. As he wrote his friend Hilbert, in 1889, [14, p.38]

> I have now progressed much further into the theory of positive definite quadratic forms, and in the case of forms with a greater number of variables it is indeed true that the situation is very different ...
>
> Perhaps the following theorem would interest you or Hurwitz (which I can prove in half a page):
>
> In a positive definite quadratic form of determinant D and n (≥ 2) variables one can always give the variables such integer values that the form is less than $nD^{1/n}$.

Hermite has here instead of the coefficient n only $(4/3)^{(n-1)/2}$, which obviously in general is a much higher bound ...

Please give my best greetings to Hurwitz as well as my best regards to our other colleagues, and I would ask you, yourself, to remember—and now and then to write to ....

Your faithful
H. Minkowski

Exercise 12. Using a calculator find the first n for which Minkowski's bound is smaller than Hermite's.

Exercise 13. Derive Minkowski's bound for the general case from the special case when $D = 1$.

Minkowski's approach is based on properties of translates of a set K by the vectors in a lattice L. Consider a lattice L that has determinant d and a set K of area A. If v is a vector, the set of points $\{v+k : k \in K\}$ is called a *translate* of K and denoted $v+K$. Assume that the translates of K by the vectors in L are disjoint. What can we then say about the area of A? Figure 8 shows the typical situation, where K is a worm-shaped set.

In Figure 8 we have shaded the part of one parallelogram that is occuped by translates of K. Call this set S. Every part of the worm appears in S. After all, each part appears in some parallelogram, and whatever appears in one parallelogram appears in all of them. Thus the area of S is A. Since S lies in a parallelogram of area d we conclude that $A \leq d$. Lemma 1 puts this observation in a form that will be more useful. Note that our reasoning applies in all dimensions.

Lemma 1. *Let L be an n-dimensional lattice of determinant d. Let K be a set in n-space of volume greater than d. Then there are distinct elements v_1 and v_2 in L such that $(v_1 + K) \cap (v_2 + K)$ is not empty.*

Exercise 14. Show that Lemma 1 does not hold if the volume of K is only d.

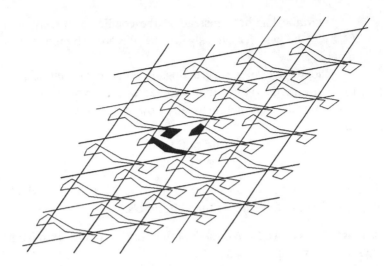

FIGURE 8

The theorem that Minkowski deduced from this lemma concerns centrally symmetric convex sets. For convenience we will restrict our attention to bounded sets. A bounded set C in n-space is *convex* if, whenever points P and Q are in C so is the entire line segment PQ. It is *centrally symmetric* if there is a point M such that for all points P in C, the point Q such that M is the midpoint of PQ is also in C. An ellipse or parallelogram is a centrally symmetric convex set, but a triangle is not.

Theorem 1. *Let C be a centrally symmetric convex set in n-space. Assume that its volume is greater than $2^n d$ and the origin is its center of symmetry. Let L be an n-dimensional lattice of determinant d. Then C contains a point of L other than the origin.*

Proof. For convenience suppose $n = 2$. Let $K = \{x/2 : x \in C\}$, which is similar to C but shrunk by a factor of 2 in all directions, hence it has area greater than d. It is shown in Figure 9.

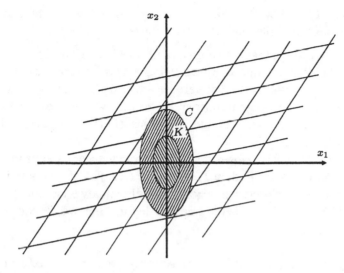

FIGURE 9

Since K has area greater than d, we conclude that two translates of K meet, that is, there are vectors $x_1, x_2 \in L$ and $k_1, k_2 \in K$,
$x_1 \neq x_2$, such that

$$x_1 + k_1 = x_2 + k_2.$$

Hence

$$x_1 - x_2 = k_2 - k_1.$$

Since K is centrally symmetric, $-k_1$ is an element K, which we denote k_3. Hence

$$x_1 - x_2 = k_2 + k_3 = 2\left(\frac{k_2 + k_3}{2}\right).$$

Since K is convex $(k_2 + k_3)/2 \in K$ and therefore $2(k_2 + k_3)/2 \in C$,
that is, $k_2 + k_3 \in C$. Hence C contains $x_1 - x_2$, a nonzero vector in
L and the proof is complete.   □

Exercise 15. Show that the theorem is not true if either of the assumptions "centrally symmetric" or "convex" is removed.

If we assume that the volume of C is $2^n d$, then Theorem 1 no longer holds. A counterexample in 2-space is provided by the lattice Z^2 and C is the square whose vertices are $(1, 1)$, $(-1, 1)$, $(-1, -1)$, and $(1, -1)$, but without its border. However, if we add the assumption that C is a "closed" set, then the conclusion of Theorem 1 still holds, as we will show.

A set C in R^n is *closed* if whenever a sequence of points in C converges to a point P, then P must be in C. A polygon without its boundary is not closed; a polygon with its boundary is closed.

The next theorem is the one that led to Minkowski's conjecture.

Theorem 2. *Let C be a bounded, centrally symmetric, closed set in n-space. Assume that its volume is $2^n d$ and the origin is its center of symmetry. Let L be an n-dimensional lattice of determinant d. Then C contains a point of L other than the origin.*

Proof. Assume that the theorem is false. Enclose C in some very large box, much larger than C. This box contains only a finite number of points of L, and let P be one of them other than the origin. We first show that the distance between P and any point in C is greater than some fixed positive number.

Let $d(P, Q)$ denote the distance between P and point Q. Assume that for each positive integer m there is a point Q_m in C such that $d(P, Q_m)$ is less than $1/m$. The sequence $Q_1, Q_2, \ldots, Q_m, \ldots$ converges to P. Since C is closed, P lies in C. This contradicts the assumption that P is not in C.

Therefore there is a positive number p such that $d(P, Q) > p$ for all points Q in C. It follows that there is a positive number p^* such that the distance between any point of C and any point of L other than the origin is greater than p^*.

Let s be a number slightly larger than 1. Let $sC = \{sQ : Q \in C\}$, which is a slight magnification of C. Indeed, choose s so that $d(sQ, Q) < p^*$ for all Q in C. Then sC has volume $s^n d$, which is greater than d, and contains no point of L other than the origin. This contradiction of Theorem 1 completes the proof.   □

Minkowski first developed these theorems for the special cases when C is a box or ball. Later he showed that they hold for any centrally symmetric convex set. The reason he wanted these theorems is illustrated by the following consequence.

Theorem 3. *A plane lattice L of determinant 1 contains an element whose length is not greater than $\sqrt{2}$.*

Proof. Let C be the square whose vertices are $(1,1), (-1,1)$, $(-1,-1)$, and $(1,-1)$. By Theorem 1 it contains an element $v \in L$, $v \neq (0,0)$. But every point of C is within a distance $\sqrt{2}$ of the origin. Thus the length of v is less than or equal to $\sqrt{2}$.   □

Exercise 16. Justify the claim Minkowski made in his letter.

Minkowski improved the bound in Theorem 3 by using balls instead of cubes.

Exercise 17. Using a disk instead of a square, prove that a plane lattice of determinant 1 contains an element whose length is not greater than $2/\sqrt{\pi}$. (Note that $2/\sqrt{\pi} \approx 1.13$, which is less than $\sqrt{2}$, the bound in Theorem 2. This is not the best possible result.)

Exercise 18. What does the previous exercise imply about positive definite quadratic forms in two variables?

In Minkowski's obituary, Hilbert [**13**, pp. v–xxi] commented on Theorem 2,

> This proof of a deep number theoretical result without analytical tools, essentially on the basis of a geometrically intuitive consideration, is a pearl of Minkowski's discoveries. In the generalization to forms of n variables Minkowski's proofs led to a more natural and far lower

upper bound than that which Hermite had found. Still
more important however than this was that the essential
idea of Minkowski's reasoning used only this property of
the ellipsoid, namely that it is a convex figure and pos-
sesses a center, which could therefore be applied to arbitrary convex figures with a center. This situation led
Minkowski to recognize for the first time that the concept
of a convex body is a fundamental idea in our science.

4. Minkowski's conjecture

Minkowski and others used Theorem 2, the "lattice point theorem,"
to obtain several fundamental results in algebraic number theory.
However to pursue this path would take us far from the route we
want to follow. It is Minkowski's use of his theorem in examining
the simultaneous approximation of several real numbers by rational
numbers that will lead us to consider tilings of space by parallel
congruent cubes.

Rational numbers with a given denominator, x_2, can approxi-
mate any number a fairly well. A glance at the number line in Figure
10 shows that there is an integer x_1 such that

$$\left| a - \frac{x_1}{x_2} \right| \leq \frac{1}{2x_2}.$$

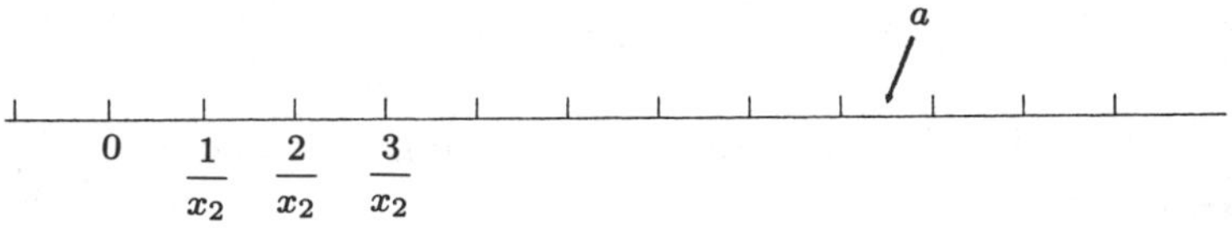

FIGURE 10

The next exercise obtains a stronger result by using Minkowski's
theorem.

Exercise 19. Let a be a real number and n a positive integer. Prove that there are integers x_1 and x_2 such that

$$\left| a - \frac{x_1}{x_2} \right| \le \frac{1}{nx_2} \qquad \text{and} \qquad x_2 \le n.$$

Suggestion: Let L be the lattice in R^2 with basis $(a, 1)$ and $(-1, 0)$. Let C be a rectangle of width $2/n$ and height $2n$.

Minkowski considered the problem of approximating two real numbers, a and b, by rational numbers with the same denominator, but not prescribing the denominator, and proved the following theorem.

Theorem 4. *Let t be a real number greater than 1 and a and b real numbers. Then there are integers x_1, x_2 and x_3, with x_3 positive, such that*

$$\left| a - \frac{x_1}{x_3} \right| \le \frac{1}{x_3 t}, \qquad \left| b - \frac{x_2}{x_3} \right| \le \frac{1}{x_3 t}, \qquad \text{and} \qquad x_3 \le t^2,$$

hence

$$\left| a - \frac{x_1}{x_3} \right| \le \frac{1}{x_3^{3/2}}, \qquad \text{and} \qquad \left| b - \frac{x_2}{x_3} \right| \le \frac{1}{x_3^{3/2}}.$$

Proof. Let $B = (b_{ij})$ be a 3 by 3 real matrix of determinant 1. The cube with vertices $(\pm 1, \pm 1, \pm 1)$ contains a nonzero vector of the lattice spanned by the rows of B. Thus the simultaneous inequalities

$$|b_{11}x_1 + b_{12}x_2 + b_{13}x_3| \le 1$$
$$|b_{21}x_1 + b_{22}x_2 + b_{23}x_3| \le 1 \qquad\qquad (1)$$
$$|b_{31}x_1 + b_{32}x_2 + b_{33}x_3| \le 1$$

have an integer solution other than the trivial one, $(0, 0, 0)$. Now let $t > 1$ be a real number and a and b be real numbers. As a special case of (1) the inequalities

$$|tx_1 + 0x_2 - \quad atx_3| \leq 1$$
$$|0x_1 + tx_2 - \quad btx_3| \leq 1$$
$$|0x_1 + 0x_2 + (1/t^2)x_3| \leq 1$$

have a nontrivial integer solution. Thus there are integers x_1, x_2, and x_3, with $x_3 > 0$, such that

$$\left|a - \frac{x_1}{x_3}\right| \leq \frac{1}{x_3 t}, \qquad \left|b - \frac{x_2}{x_3}\right| \leq \frac{1}{x_3 t}, \qquad x_3 \leq t^2,$$

and consequently

$$\left|a - \frac{x_1}{x_3}\right| \leq \frac{1}{x_3^{3/2}}, \qquad \left|b - \frac{x_2}{x_3}\right| \leq \frac{1}{x_3^{3/2}}. \qquad \square \qquad (2)$$

Exercise 20.   State and prove the analog of (2) for three real numbers a, b, and c.

Minkowski showed that the inequalities in (1) could not always be strict. His counterexample was

$$|x_1 + b_{12}x_2 + b_{13}x_3| < 1$$
$$|0x_1 + \quad x_2 + b_{23}x_3| < 1 \qquad (3)$$
$$|0x_1 + \quad 0x_2 + \quad x_3| < 1.$$

(Note that the determinant of B is 1.)

Any integer solution of (3) has $x_3 = 0$, hence $x_2 = 0$, and finally $x_1 = 0$.

The assertion that the inequalities

$$|b_{11}x_1 + b_{12}x_2 + b_{13}x_3| < 1$$
$$|b_{21}x_1 + b_{22}x_2 + b_{23}x_3| < 1 \qquad (4)$$
$$|b_{31}x_1 + b_{32}x_2 + b_{33}x_3| < 1$$

have only the trivial solution is equivalent to saying that the interior of the cube with vertices $(\pm 1, \pm 1, \pm 1)$ contains only one point of the lattice spanned by

$$v_1 = (b_{11}, b_{21}, b_{31}), \quad v_2 = (b_{12}, b_{22}, b_{32}), \quad v_3 = (b_{13}, b_{23}, b_{33}),$$

namely, the origin. This, in turn, is equivalent to the assertion that the translates of the interior of the closed cube K with vertices $(\pm 1/2, \pm 1/2, \pm 1/2)$ by the vectors of the lattice spanned by v_1, v_2 and v_3 are disjoint. Now, K has volume 1. Thus the union of the translates of K is all of 3-space.

Exercise 21. Why is the union of the cubes all of 3-space?

This is where the concept of tiling space enters the picture.

We shall define a "tiling" in sufficient generality to cover all the cases met in the following chapters.

Let $S_1, S_2, \ldots, S_k, \ldots$ be a family of subsets of a set S in n-space such that each S_k is an n-dimensional polyhedron or the finite union of such polyhedra. In Chapters 1 to 4 S_k is an n-dimensional cube or the finite union of such cubes. In Chapters 5 and 6 S_k is a triangle. The family is a *packing* of S if the interiors of the S_k's are disjoint. The family is a *covering* of S if the union of the S_k's is S. A family that is both a packing and a covering of S is called a *tiling* of S. If each S_k is congruent to fixed set T and the family of S_k's packs (covers, tiles) S, we say that T *packs* (*covers, tiles*) S. More specifically, if S_k is a translate of a fixed set T, we say that T packs (covers, tiles) S *by translates*. If the context makes it clear that we are considering only translates, we will usually omit the phrase "by translates." Chapters 1, 2, 3, 5, and 6 concern only tiling, while Chapter 4 treats packing and covering. In Chapters 1, 2, 3, and 4 S is n-space, while in Chapters 5 and 6 it is a polygon.

As we have just seen, Minkowski was led by an algebraic question to consider tilings of space by translates of a unit cube.

What does a lattice of translates of a unit cube in 3-space look like? To answer this question, first consider a lattice of translates of a unit square K that tiles 2-space.

For convenience, let K be the square whose vertices are $(0, 0)$, $(1, 0)$, $(1, 1)$, and $(0, 1)$. If no square shares a complete edge with K, then the squares that lie along the right-hand edge or top edge of K are arranged as in Figure 11 or Figure 12.

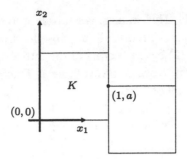

FIGURE 11

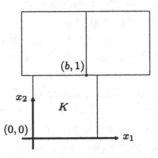

FIGURE 12

In the first case two squares share a complete common horizontal edge. In other words, the lattice contains the unit vector $(0, 1)$. In the second case, the lattice contains the vector $(1, 0)$.

In the first case, the tiling is made up of vertical strips staggered uniformly, as in Figure 13. In the second case it is made up of horizontal strips.

The lattice indicated in Figure 13 has a basis of the form $(1, a)$ and $(0, 1)$, for some number a. Thus a typical vector is $x_1(1, a) + x_2(0, 1)$. Since no such vector lies in the square whose vertices are $(\pm 1, \pm 1)$, the inequalities

$$|x_1 + 0x_2| < 1$$
$$|ax_1 + x_2| < 1 \tag{5}$$

have only one integer solution, $(0, 0)$.

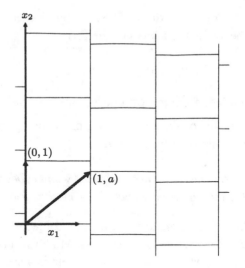

FIGURE 13

Exercise 22. Assume that $B = (b_{ij})$ has determinant 1 and the inequalities

$$|b_{11}x_1 + b_{12}x_2| < 1$$
$$|b_{21}x_1 + b_{22}x_2| < 1$$

have only one integer solution, $(0,0)$. Show that some row of B consists of integers and they are relatively prime. (Hint: Translate the problem into one on squares.)

With the aid of a few sugar cubes or dice it is not hard to show that if translates of a unit cube tile 3-space, then some pair share a complete 2-dimensional face. In fact, by the time that you surround a cube with other cubes, you will find that two share a complete face. Thus a lattice tiling of 3-space by parallel congruent cubes consists of infinite tubes. These tubes, in turn, form slabs, and the slabs fill up 3-space.

Exercise 23.   Consider a tiling of 3-space by a lattice of unit cubes. Show that the lattice has a basis of the form

$$(1, 0, 0), \ (x_1, 1, 0), \ (x_2, x_3, 1),$$

or a basis obtained from this by a permutation of the coordinates.

This was Minkowski's conjecture, made in algebraic form in 1896 and in geometric form in 1907.

Conjecture.   *In a lattice tiling of n-space by unit cubes there must be a pair of cubes that share a complete $(n-1)$-dimensional face.*

Actually, he never made this a specific conjecture. In his book, *Geometrie der Zahlen*, published in 1896, Minkowski [11] wrote on p. 105, "I plan to give a proof of this theorem in a special article in connection with detailed arithmetic investigations of n linear forms, exactly like I will carry out here [in Section 45] for the case $n = 2$."

The argument in Section 45 did not easily generalize to all dimensions, and no "special article" appeared. In his *Diophantische Approximationen* [12], published in 1907, we find on page 28, "I assume and would like to pose it as a problem." He settled the case $n = 3$ on pages 67–74, translated this case into a property of a lattice tiling of 3-space by cubes, and included the diagram shown in Figure 14 that suggests the general appearance of a lattice of cubes. However, his argument for the case $n = 3$ again failed to generalize to higher dimensions.

In spite of its seeming simplicity, Minkowski's conjecture reached the venerable age of 45 years before it was settled by Hajós in 1941 [5].

We will call two n-dimensional cubes whose intersection is a complete $(n-1)$-dimensional face *twins*. Schmidt in 1933 [19] observed that to prove Minkowski's conjecture it is enough to prove it for the case of a rational lattice cube tiling, that is, for a lattice cube tiling in which the lattice vectors have only rational coordinates. Here we tacitly assume that the cubes are unit cubes and they are parallel to the coordinate system, since the rationality depends on

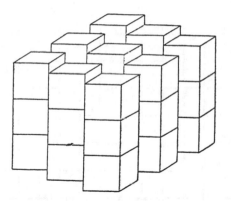

FIGURE 14

the choice of coordinate system. We will justify this reduction in the next chapter.

5. The group theoretic version of Minkowski's conjecture

It was a milestone when Hajós in 1938 translated Minkowski's conjecture into an equivalent conjecture about finite abelian groups. Three years later this led him to his proof of the conjecture. We will illustrate the connection between the geometry of lattice cube tilings and abelian groups by describing it in 2-space.

Consider a lattice family of translates of a unit square, whose vertices are $(0,0)$, $(1,0)$, $(1,1)$ and $(0,1)$, which need not form a tiling. Let L be the vectors of the lattice. (They are the bottom left corners of the squares.) We restrict our attention only to rational lattices. Thus if l_1 and l_2 are basis vectors of L, then we may assume that

$$l_1 = \left(\frac{a_{11}}{b_{11}}, \frac{a_{12}}{b_{12}}\right), \qquad l_2 = \left(\frac{a_{21}}{b_{21}}, \frac{a_{22}}{b_{22}}\right),$$

where a_{ij}, b_{ij} are integers and $b_{ij} > 0$. Let r_1 be a positive common multiple of b_{11} and b_{21} and similarly let r_2 be a positive common

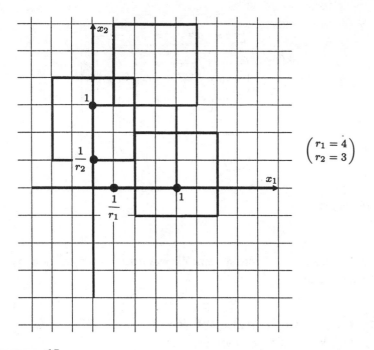

FIGURE 15

multiple of b_{12} and b_{22}. The straight lines perpendicular to the first axis going through the points whose first coordinates are integer multiples of $1/r_1$ and the straight lines perpendicular to the second axis going through the points whose second coordinates are integer multiples of $1/r_2$ cut the plane into rectangles of dimensions $1/r_1$ by $1/r_2$, as shown in Figure 15.

Each square of the family is cut into $r_1 r_2$ rectangles whose dimensions are $1/r_1$ by $1/r_2$. The small rectangles form a lattice tiling of 2-space by the lattice L' consisting of their bottom left corners. This lattice has a basis

$$l_1' = (1/r_1, 0), \qquad l_2' = (0, 1/r_2).$$

Now, L is a sublattice of L', or in other words L is a subgroup of L'. The main algebraic tool we need is the factor group $G = L'/L$, whose elements are the cosets modulo L, or, geometrically speaking, the translates of L. The coset L is the identity element e of G. Let a_1 and a_2 be the cosets containing the basis vectors l_1' and l_2' of L'. If the interiors of two members of the family of squares overlap, that is, if one of the small rectangles is covered by two squares, then an element l' of L' can be represented in two distinct ways in the form

$$l' = l + x_1 l_1' + x_2 l_2',$$

where $l \in L$, $0 \leq x_1 \leq r_1 - 1$, $0 \leq x_2 \leq r_2 - 1$. On the other hand, if there is a rectangle that is not covered by any square of the family then there is an element l' of L' which cannot be represented in the above form. Using the factor group G written multiplicatively we may say that if the elements

$$a_1^{x_1} a_2^{x_2}, \qquad 0 \leq x_1 \leq r_1 - 1, \quad 0 \leq x_2 \leq r_2 - 1 \qquad (6)$$

are distinct then the lattice of squares is a packing of 2-space. If these elements represent all the elements of G then the lattice of squares is a covering of 2-space. If these elements represent all elements of G without any repetition then the lattice of squares is a tiling. *Thus by means of G we can state whether the corresponding lattice of squares is a tiling.*

Next we translate into the language of the group G the assertion that the lattice of squares has no twins.

Two squares form a pair of twins if the lattice elements that describe their position differ by $(1, 0)$ or $(0, 1)$. Thus, if twins are present, $(1, 0)$ or $(0, 1)$ belongs to the lattice L. In other words, $r_1 l_1' \in L$ or $r_2 l_2' \in L$. In the factor group G we then have

$$a_1^{r_1} = e \qquad \text{or} \qquad a_2^{r_2} = e. \qquad (7)$$

That twins are present in the lattice square tiling can be formulated in terms of G as follows: If every element in the abelian group G can be represented uniquely in the form (6), then (7) follows.

Similar reasoning in n-space gives us Hajós's translation of Minkowski's conjecture into an equivalent conjecture about finite abelian groups.

Hajós's version of Minkowski's conjecture. *Let G be a finite abelian group. If $a_1, a_2, \ldots, a_n$ are elements of G and $r_1, r_2, \ldots, r_n$ are positive integers such that each element of G is uniquely expressible in the form*

$$a_1^{x_1} \cdots a_n^{x_n}, \qquad 0 \le x_1 \le r_1 - 1, \ldots, 0 \le x_n \le r_n - 1,$$

then $a_i^{r_i} = e$ for some i, $1 \le i \le n$.

Since we have verified Minkowski's conjecture for 1-space and 2-space, and sugar cubes quickly settle it in 3-space, we know that Hajós's version holds for $n \le 3$.

Exercise 24. Show that if Minkowski's conjecture is true when $r_1 = \cdots = r_n$, then it is true in general. (Hint: Go back to the geometry.)

6. More about the algebraic version of Minkowski's conjecture

Let G be an abelian group and let $A_1, \ldots, A_n$ be subsets of G. If each element g of G is uniquely expressible in the form

$$g = a_1 \cdots a_n, \qquad a_1 \in A_1, \ldots, a_n \in A_n,$$

then we say that G is *factored* by the subsets $A_1, \ldots, A_n$ and that the product $A_1 \cdots A_n$ is a *factorization* of G. (Note that if G is finite then $|G| = |A_1| \cdots |A_n|$.) This concept is a natural extension of the concept of direct product of subgroups and we will write $G = A_1 A_2 \cdots A_n$. We will assume throughout that each A_i is normalized to contain the identity element of G.

Let a be an element of the abelian group G and r a positive integer. The subset consisting of the elements

$$e, a, a^2, \ldots, a^{r-1}$$

will be called a *cyclic subset*. To avoid trivial cases we suppose that $a \neq e$, $r \geq 2$ and that the order of a is at least r, so that the elements $e, a, a^2, \ldots, a^{r-1}$ are distinct. If the order of a is r then this cyclic subset coincides with the cyclic group generated by a. A cyclic subset can be thought of as a "front end" of a cyclic group.

Hajós's version now reads: *In any factorization of a finite abelian group by cyclic subsets at least one of the factors is a subgroup.*

The conjecture in this form could be thought of as the dual of the fundamental theorem of finite abelian groups. Indeed, the latter states that there is a factorization of G by cyclic subsets in which all the factors are subgroups. The algebraic version of Minkowski's conjecture, on the other hand, claims that in each factorization of G by cyclic subsets there is at least one subgroup among the factors.

The next two exercises show that in Hajós's version we may assume that the cyclic subsets have prime orders.

Exercise 25.
(a) Verify that if $r \geq 2$ and $s \geq 2$, then the cyclic subset

$$A = \{e, a, a^2, \ldots, a^{rs-1}\}$$

of cardinality rs can be factored into the smaller cyclic subsets

$$B = \{e, a, a^2, \ldots, a^{r-1}\} \quad \text{and} \quad C = \{e, a^r, a^{2r}, \ldots, a^{(s-1)r}\}$$

of cardinalities r and s respectively.
(b) Show that A can also be factored into the cyclic subsets

$$B' = \{e, a, a^2, \ldots, a^{s-1}\} \quad \text{and} \quad C' = \{e, a^s, a^{2s}, \ldots, a^{(r-1)s}\}$$

and so the factorization is not always unique.

Exercise 26. (This continues Exercise 25.)
(a) Prove that B cannot be a subgroup.
(b) Prove that A is a subgroup of G if and only if C is a subgroup of G.

By Exercise 25 every cyclic subset can be factored into cyclic subsets of prime cardinalities. By Exercise 26 if the original cyclic subset is not a subgroup then neither are the new ones; and if none of the new ones is a subgroup then neither is the original. Thus we may suppose that in Hajós's version all the cyclic subsets are of prime orders. But now the number of the factors need no longer correspond to the dimension of the original cube tiling. As we mentioned earlier, Hajós settled Minkowski's conjecture in 1941 using this approach.

7. Related work

In 1930 Keller [6] suggested that the "lattice" condition in Minkowski's problem is irrelevant, conjecturing that in a tiling of n-space by parallel unit cubes there must be twins. Perron in 1940 [16] verified this conjecture for $n \leq 6$.

If this conjecture were true it would imply Hajós's theorem, and show that Minkowski's theorem was really geometric, not algebraic. However, in 1992 Lagarias and Shor [9], using an approach of Corrádi and Szabó [2], showed that Keller's conjecture is false in all dimensions greater than or equal to 10. (Dimensions 7, 8, and 9 are not settled.)

Similarly, Rédei wondered whether the sets of prime orders in Hajós's theorem had to be cyclic. In 1965 he showed [17] that the assumption could be removed, proving the following theorem.

Rédei's Theorem. *Let G be a finite abelian group and $A_1, A_2, \ldots, A_n$ be normalized subsets of G of prime orders. Assume that $G = A_1 A_2 \cdots A_n$ is a factorization. Then at least one of the sets A_i is a subgroup.*

We present a simplified version of Rédei's proof in Chapter 7. We delay it till then, even though it is elementary, since it involves a lengthy sequence of lemmas. However, we will make use of it in the next chapter to obtain a surprising property of certain sets that are made up of a prime number of cubes.

Exercise 27. How would Rédei's theorem read if we remove the assumption that each A_i is normalized?

In 1936 Furtwängler [3] generalized Minkowski's twin problem to a problem about multiple tilings, which we describe.

Consider a lattice family of cubes that are translates of each other. Suppose that every point of space belongs to finitely many cubes of the family. Further, suppose that any point that is not on the boundary of any cube lies in precisely k cubes of the system. This cube system is called a k-*fold tiling* and its multiplicity is k. Furtwängler conjectured that in such a k-fold lattice tiling there would be twins.

Exercise 28. Show that Furtwängler's conjecture is equivalent to this variant of Hajós's version: Let G be a finite abelian group and $A_1, A_2, \ldots, A_n$ cyclic subsets of G such that each element g of G is representable exactly k ways in the form $g = a_1 a_2 \cdots a_n$, where a_i is in A_i. (This is called a k-*factorization* of G by subsets $A_1, A_2, \ldots, A_n$.) Then at least one of the A_i is a subgroup of G.

Furtwängler proved his conjecture for $n \leq 3$. In 1938 Hajós, using the algebraic version of the conjecture, exhibited a 9-fold cube tiling in 4-space that has no twins. In 1979 R. M. Robinson [18] characterized all pairs (k, n) for which there exists a k-fold n-dimensional lattice cube tiling without twins. His result is:

If $n \leq 3$, there is no such k.

If $n = 4$, k is any integer divisible by the square of an
      odd prime.

If $n = 5$, then $k = 3$ or $k \geq 5$.

If $n \geq 6$, then each $k \geq 2$ is possible.

Exercise 29. Show that Robinson's result implies Hajós's theorem.

The next exercise refutes Furtwängler's conjecture in 6-space.

Exercise 30. Let G be the direct product of two cyclic groups of order 4 and let x and y be a basis for G, hence $x^4 = e = y^4$.

(a) Show that none of the cyclic subsets

$$A_1 = \{e, y\}, \qquad A_2 = \{e, x^2 y\}, \qquad A_3 = \{e, x\},$$
$$A_4 = \{e, xy\}, \qquad A_5 = \{e, xy^2\}, \qquad A_6 = \{e, xy^3\}$$

is a subgroup of G.

(b) Verify that the product $A_1 A_2 A_3 A_4 A_5 A_6$ is a 4-factorization of G.

(c) What does (b) tell us about multiple cube tilings?

The next exercise gives another counterexample.

Exercise 31. Let G be the direct product of cyclic groups of orders 2, 2, and 3 and suppose that u, v, and w form a basis for G such that $u^2 = v^2 = w^3 = e$.

(a) Show that none of the following cyclic subsets is a subgroup of G:

$$A_1 = \{e, vw, (vw)^2\}, \qquad A_2 = \{e, uw, (uw)^2\},$$
$$A_3 = \{e, uvw, (uvw)^2\}, \qquad A_4 = \{e, vw\}, \qquad A_5 = \{e, uw\}.$$

(b) Verify that the product $A_1 A_2 A_3 A_4 A_5$ is a 9-factorization of G.

(c) What does (b) tells us about multiple cube tilings?

Another path leads off from quadratic forms. Consider a positive definite quadratic form $x^T A x$ that assumes only integer values when the vector x has only integer coordinates. Represent A in the form $B^T B$. Let the rows of B be $v_1, v_2, \ldots, v_n$. Then the square of the length of every vector in the lattice whose basis is $v_1, v_2, \ldots, v_n$ is an integer. Such a lattice is called an *integral lattice*. This should be distinguished from the "integer lattice" Z^n, which is a special case of an integral lattice.

Exercise 32. Show that the vectors $v_1, v_2, \ldots, v_n$ generate an integral lattice if and only if $v_i \cdot v_i$ and $2v_i \cdot v_j$ are integers for all $1 \le i$, $j \le n$.

For an integral lattice L of determinant 1, consider the minimum value of the square of the lengths of the nonzero vectors of L.

Denote this by $\mu(L)$. Let μ_n be the maximum value of $\mu(L)$ for all n-dimensional integral lattices L of determinant 1.

Exercise 33. Using Minkowski's bounds on the shortest vector in any lattice, show that $\mu_1 = 1$ and $\mu_2 = 1$.

Conway and Sloane [1] proved that for all sufficiently large n, $\mu_n \leq (n+6)/10$ and included this table of values:

n	1–7	8	9	10	11	12	13	14–22	23	24	25	26–31	32	33
μ_n	1	2	1	1	1	2	1	2	3	4	2	3	4	3

In 1982, Lenstra, Lenstra, and Lovász [10] developed an efficient algorithm for finding short vectors in a lattice and applied it to the factoring of polynomials with integer coefficients.

Lagarias and Odlyzko [8] in 1985 applied it to public-key knapsack cryptography. Odlyzko and te Riele in 1985 [15] used that algorithm to show that a conjecture made by Mertens in 1897 is false. (Had the conjecture been true, it would have implied the Riemann hypothesis.) Mertens' conjecture concerns the prime factorizations of those integers which are not divisible by a square larger than 1, the "square-free" integers. The prime factorizations of such integers have no repeated factor. For a positive number x let

$$M(x) = \left(\begin{array}{l} \text{number of square-free integers less than or} \\ \text{equal to } x \text{ that have an even number of prime} \\ \text{factors} \end{array} \right)$$

$$- \left(\begin{array}{l} \text{number of square-free integers less than or} \\ \text{equal to } x \text{ that have an odd number of prime} \\ \text{factors} \end{array} \right).$$

For instance, the square-free integers less than or equal to 10 are 1, 2, 3, 5, 6, 7, and 10. Of these, 1, 6, and 10 have an even number of prime factors and 2, 3, 5, and 7 have an odd number of prime factors. Thus $M(10) = 3 - 4 = -1$.

Exercise 34. Compute $M(x)$ for $x = 25$ and $x = 36$ and compare $|M(x)|$ to $\sqrt{x}$.

On the basis of a tabulation of $M(x)$ for x up to 10,000, Mertens conjectured that for $x > 1$

$$|M(x)| < \sqrt{x}.$$

Odlyzko and te Riele showed that $M(x) > 1.06\sqrt{x}$ and $M(x) < -1.009\sqrt{x}$ for an infinite number of integers x, but confessed that, "Our proof is indirect and does not produce any single value of x for which $|M(x)| > \sqrt{x}$. In fact, we suspect that there are no counterexamples to Mertens' conjecture for $x \leq 10^{20}$ or even 10^{30}," and gave reasons for this belief. For all large values of x for which $M(x)$ has been evaluated, specifically, up to 7.8×10^9, $|M(x)| < 0.6\sqrt{x}$, illustrating the saying, "Any integer you can test is too small."

This chapter showed how even a question about tiling space with cubes takes us quickly from the realm of geometry into algebra and number theory. Hajós used not only the group G but leaned heavily on the group ring formed on G. As we will see in Chapter 7, the proof of Rédei's generalization also employs the group ring, but only briefly, and brings in the character group of G and cyclotomic fields as well.

This chapter also illustrates how intricate is the web of mathematics, for we have seen such varied areas as quadratic forms, lattices, convex sets, approximation by rationals, tiling by cubes, packing spheres, factorization of abelian groups by subsets, factoring polynomials, and number theory overlapping each other. Clearly any barrier inserted between one part of mathematics and another would be quite artificial. Subsequent chapters will provide far more evidence for this assertion.

References

1. J. H. Conway and N. J. A. Sloane, A new upper bound for the minimum of integral lattice of determinant 1, *Bull. Amer. Math. Soc.* **23** (1990), 383–387.

2. K. Corrádi and S. Szabó, A combinatorial approach for Keller's conjecture, *Periodica Mat. Hung.* **21** (1990), 95–100.

3. Ph. Furtwängler, Über Gitter konstanter Dichte, *Monatsh. Math. Phys.* **43** (1936), 281–288.

4. C. F. Gauss, *Werke* Vol. 2, König. Gesellschaft der Wiss. Göttingen, 1876.

5. G. Hajós, Über einfache und mehrfache Bedeckung des n-dimensionalen Raumes mit einem Würfelgitter, *Math. Zeit.* **47** (1942), 427–467.

6. O. H. Keller, Über die lückenlose Einfüllung des Raumes mit Würfeln, *J. Reine Angew. Math.* **163** (1930), 231–248.

7. ——, Ein Satz über die lückenlose Erfüllung des 5- und 6-dimensional Raumes mit Würfeln, *J. Reine Angew. Math.* **177** (1937), 61–64.

8. J. F. Lagarias and A. M. Odlyzko, Solving low-density subset sum problems, *JACM* **32** (1985), 229–246.

9. J. F. Lagarias and P. W. Shor, Keller's cube-tiling conjecture is false in high dimensions, *Bull. Amer. Math. Soc.* **27** (1992), 279–283.

10. A. K. Lenstra, H. W. Lenstra, and L. Lovász, Factoring polynomials with rational coefficients, *Math. Ann.* **261** (1982), 515–534.

11. H. Minkowski, *Geometrie der Zahlen,* Teubner, Leipzig, 1896.

12. ——, *Diophantische Approximationen,* Teubner, Leipzig, 1907 (reprint: Physica-Verlag, Würzberg, 1961).

13. ——, *Gesammelte Abhandlungen von Hermann Minkowski,* Chelsea, New York, 1967.

14. ——, *Briefe an David Hilbert,* Springer-Verlag, New York 1973.

15. A. M. Odlyzko and H. J. J. te Riele, Disproof of Mertens' conjecture, *J. Reine Angew. Math.* **357** (1985), 138–160.

16. O. Perron, Modulartige lückenlose Ausfüllung des R^n mit kongruenten Würfeln I., II., *Math. Ann.* **117** (1940), 415–447, (1941), 609–658.

17. L. Rédei, Die neue Theorie der endlichen abelschen Gruppen und Verallgemeinerung des Hauptsatzes von Hajós, *Acta Math. Acad. Sci. Hung.* **16** (1965), 329–373.

18. R. M. Robinson, Multiple tilings of n-dimensional space by unit cubes, *Math. Zeit.* **166** (1979), 225–264.

19. T. Schmidt, Über die Zerlegung des n-dimensionalen Raumes in gitterförmig angeordnete Würfeln, *Schr. math. Semin. u. Inst. angew. Math. Univ. Berlin* **1** (1993), 186–212.

20. S. K. Stein, Algebraic tiling, *Amer. Math. Monthly* **81** (1974), 445–462.

21. S. Szabó, A reduction of Keller's conjecture, *Periodica Math. Hung.* **17** (1986), 265–277.

22. M. F. Woepcke, Recherches sur plusieurs ouvrages de Léonard de Pise decouverts et publiés par M. le prince Balthasar Boncompagni et sur les rapports que existent entre les ouvrages et les travaux mathématiques des Arabes, *Atti della Academia dei Lincei* **14** (1861), 301–324.

Chapter 2
Cubical Clusters

In Chapter 1 we were concerned with the way translates of a single cube fit together to tile space. In this chapter we examine tilings by translates of a finite collection of cubes, which we will call "clusters." Chapters 3 and 4 will treat a special family of clusters that exists in all dimensions. Before we can state the main results of this chapter, we need some definitions.

As in Chapter 1, we assume a fixed coordinate system. We continue to identify each unit cube whose edges are parallel to the axes with its vertex that has the smallest coordinates. An n-dimensional *cluster* C is the finite union of unit cubes whose edges are parallel to the axes and which have integer coordinates. A cluster is not necessarily connected

Let C be a fixed cluster in n-space and assume that L is a set of vectors in n-space such that the set of translates $\{v + C : v \in L\}$ tile n-space. (For a given cluster there may be no such lattice.) If all the coordinates of all the vectors in L are integers (rational numbers), we speak of an *integer tiling* (*rational tiling*) by C, or simply a Z-*tiling* (Q-*tiling*). If L is a lattice we speak of *lattice tiling* by C. Combining the two notions, we speak of a Z-*lattice tiling* and a Q-*lattice tiling*.

We will prove the following theorems, all of which concern tilings by translates of a cluster. The main technique we will use goes back to Schmidt in 1933 [7].

Theorem 1.   *If there is a lattice tiling by cluster C, then there is a Q-lattice tiling by C.*

Theorem 2.   *If there is a lattice tiling by cluster C that is not a Z-tiling, then there is a Q-lattice tiling by C that is not a Z-tiling.*

Theorem 3.   *Let C be the cluster consisting of a single cube. If there is a lattice tiling by C without twins, then there is a Q-lattice tiling by C without twins.*

Theorem 4.   *If there is a tiling by cluster C, then there is a Z-tiling by C.*

Theorem 3 is the key that makes Minkowski's conjecture accessible by algebraic techniques, as we saw in Chapter 1. We will combine Theorem 2 and Rédei's theorem to prove the following theorem about clusters that consist of a prime number of cubes.

Theorem 5.   *If cluster C has a prime number of cubes and contains the $n + 1$ cubes with centers at*

$$(0,\dots,0),\ (1,0,\dots,0),\dots,(0,\dots,0,1),$$

then any lattice tiling of n-space by C is a Z-lattice tiling.

This last assertion is quite surprising. If we delete either one of the assumptions, it is false. For instance, the cluster consisting of the two unit squares whose centers are $(0,0)$ and $(2,0)$ Q-lattice tiles the plane but does not Z-lattice tile it. Also the cluster consisting of the four unit squares corresponding to $(0,0)$, $(1,0)$, $(1,1)$ and $(0,1)$ lattice tiles the plane in such a way that not all the vectors have integer coordinates.

Exercise 1.   Check that the two clusters mentioned tile as claimed.

After proving these theorems we turn our attention to the problem of filling a box with bricks. This is a tiling problem in which we wish to tile a particular bounded region with congruent copies of a very simple cluster. In this case we allow rotations of the clusters. The solution illustrates one of the simpler algebraic techniques for analyzing tiling problems.

1. Reductions

The method for altering a tiling to one with simpler translating vectors rests on a certain equivalence relation defined on the set of translation vectors. It turns out that the clusters that correspond to an equivalence class form a cylinder. This cylinder consists of a union of lines parallel to an axis, and can therefore be slid back and forth freely parallel to the axis without disturbing the other equivalence classes.

Consider two n-dimensional unit cubes in n-space corresponding to the coordinate vectors $(a_1, \ldots, a_n)$ and $(b_1, \ldots, b_n)$. If $|a_1 - b_1| = 1, |a_2 - b_2| < 1, \ldots, |a_n - b_n| < 1$ we call the two cubes *adjacent*. Geometrically speaking, the two cubes are adjacent if their intersection is an $(n-1)$-dimensional subset that is part of a face of both cubes perpendicular to the first coordinate axis. Figures 1 (a) and (b) exhibit nonadjacent cubes in 2-space and Figures 1 (c) and (d) exhibit adjacent cubes in 2-space.

Exercise 2. Verify that the geometrical and arithmetical descriptions of adjacency are equivalent.

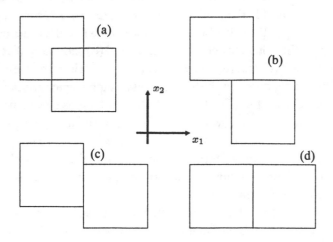

FIGURE 1

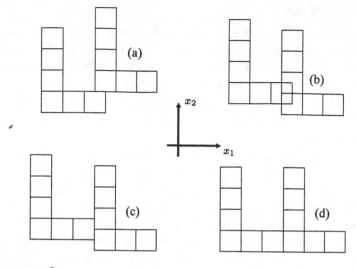

FIGURE 2

Consider a tiling of n-space by a cluster C. For convenience assume that C contains the unit cube with coordinates $(0, \ldots, 0)$. Call two clusters in this tiling *adjacent* if there are two adjacent cubes such that one of the cubes belongs to one cluster and the other cube belongs to the other cluster. Figures 2 (a) and (b) show nonadjacent and Figures 2 (c) and (d) show adjacent 2-dimensional clusters.

Adjacency between clusters induces an equivalence relation on them, as follows. Let C and C' be clusters. If there is a sequence of clusters $C_1, C_2, \ldots, C_m$, where $C = C_1, C_m = C'$ and C_i is adjacent to $C_{i+1}, 1 \le i \le m-1$, we say that C is *equivalent* to C'. Though we have defined the equivalence relation relative to the x_1-axis, clearly it could be defined relative to each of the axes. Figure 3 illustrates equivalent clusters in 2-space.

If L is the set of translating vectors for a tiling by C, we call two vectors l_1 and l_2 in L equivalent if their associated sets $l_1 + C$ and $l_2 + C$ are equivalent.

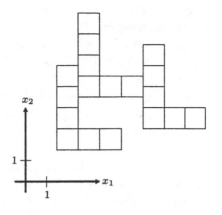

FIGURE 3

Exercise 3. Let C be the cluster consisting of just one square in R^2. Sketch the equivalence classes relative to each axis for the tiling by C when the translating vectors are

(a) $\{(m, n): m, n \in Z\}$.

(b) $\{m(1, 1/2) + n(0, 1): m, n \in Z\}$.

Exercise 4. Let C be the cluster consisting of the two squares corresponding to $(0, 0)$ and $(0, 2)$. Sketch the equivalence classes relative to each axis for the tiling by C when the translating vectors are

(a) $\{m(1/2, 1) + n(2, 0): m, n \in Z\}$

(b) $\{m(1, 0) + n(0, 4): m, n \in Z\} \cup$
    $\{m(1, 0) + n(0, 4) + (0, 1): m, n \in Z\}$.

To discover the geometric meaning of an equivalence class imagine that adjacent clusters are glued to each other along their shared surface perpendicular to the first coordinate axis. Then if we shift one of them parallel to the first axis, the entire set of clusters equivalent to it is forced to move parallel to that axis the same amount. The different equivalence classes in the tiling would slide independently of each other, for each is a cylinder whose generator is parallel to the first axis. Note that any such shift parallel to the first axis produces another tiling by C.

Now assume that the set of translating vectors, L, is a lattice, hence an abelian group, whose identity element $(0, 0, \ldots, 0)$ we denote 0. If l_1 and l_2 are in L and are equivalent, write $l_1 \sim l_2$. Note that if l_3 is in L, then $(l_1 + l_3) \sim (l_2 + l_3)$. It follows that the set $K = \{l: l \sim 0\}$ is a sublattice of L and the equivalence classes are translates of K. This is a special case of the following lemma.

Lemma 1. *Let $\sim$ be an equivalence relation on the abelian group G. Assume that if x, y, and z are in G and $x \sim y$, then $(x + z) \sim (y + z)$. Then $K = \{x: x \sim 0\}$ is a subgroup of G and the equivalence classes are cosets of K.*

Exercise 5.   Prove Lemma 1.

As a consequence of this lemma, the equivalence class K we obtained from the lattice tiling is not merely a cylinder, but is also a subgroup of L. Note that the first coordinate of any vector in K is an integer. We use the symbol K to denote both the set of vectors equivalent to $(0, \ldots, 0)$ and the union of the clusters $v + C$, $v \in K$. The context will show whether K denotes a subset of L, an algebraic object, or a union of clusters, a geometric object. Similarly, each coset of K is viewed as both an element of the factor group L/K and as a cylinder.

We shall exploit our ability to slide cylinders freely parallel to the axis relative to which they were defined to prove the reduction theorems. In our arguments e_i denotes the unit vector along the positive ith axis.

Theorem 1.   *If there is a lattice tiling by cluster C, then there is a Q-lattice tiling by C.*

Proof.   Let L be the lattice of the given tiling. Assume that some vector in L has an irrational coordinate. For convenience, assume the vector has an irrational first coordinate, that is, has the form $a_1 e_1 + \cdots + a_n e_n$, where a_1 is irrational.

Introduce the equivalence relation relative to the first axis just described and let K be the equivalence class containing $(0, \ldots, 0)$.

Introduce a second equivalence relation on L, setting two of its vectors equivalent if their first coordinates differ by a rational number. Let M be the equivalence class containing $(0, \ldots, 0)$. Note that $K \subset M$ and M is a sublattice of L but is not all of L.

It is important that M is not merely a sublattice of L but a direct summand of L, that is, there is another sublattice, T, in L such that $T \oplus M = L$. To show that M is a direct summand, we check that if $l \in L$ and n is a nonzero integer, and $nl \in M$, then $l \in M$. (This criterion is established in Appendix A.) But if nl is in M, its first coordinate is rational and therefore the first coordinate of l itself is rational, hence l is in M.

Since M is a direct summand of L there is a sublattice T of L such that $L = T \oplus M$. Let $t_1, \ldots, t_r$ be a basis for the lattice T and $t_{r+1}, \ldots, t_n$ be a basis for M. Thus $t_1, \ldots, t_n$ is a basis for L and the first coordinate of each of the vectors $t_1, \ldots, t_r$ is irrational, while the first coordinate of each of the vectors $t_{r+1}, \ldots, t_n$ is rational.

Replace $t_1, \ldots, t_r$ by

$$t_1' = t_1 + b_1 e_1, \ldots, t_r' = t_r + b_r e_1,$$

where $b_1, \ldots, b_r$ are real numbers to be chosen in a moment. Let $t_i' = t_i$ for each i, $r + 1 \leq i \leq n$. Then replace the vector $x_1 t_1 + \cdots + x_n t_n$ in L by $x_1 t_1' + \cdots + x_n t_n'$. Let L' be the lattice generated by the vectors $t_1', \ldots, t_n'$.

Since

$$x_1 t_1' + \cdots + x_n t_n' = x_1(t_1 + b_1 e_1) + \cdots + x_r(t_r + b_r e_1)$$

$$+ x_{r+1} t_{r+1} + \cdots + x_n t_n$$

$$= x_1 t_1 + \cdots + x_n t_n + (x_1 b_1 + \cdots + x_r b_r) e_1,$$

each vector in L' is obtained from a vector in L by adding a vector parallel to the first axis. Note that since K is a subset of M, its vectors remain unchanged. More generally, if $v_1 = x_1 t_1 + \cdots + x_n t_n$ and $v_2 = y_1 t_1 + \cdots + y_n t_n$ are in the same coset of K, they are changed by the same amount. Indeed, if $v_1 - v_2 \in K$, then $v_1 - v_2 \in$

M, hence $x_i - y_i = 0$, or $x_i = y_i$ for each i, $1 \leq i \leq r$. Thus, since all vectors in a cylinder $v + K$ are changed by the same amount, the cylinder, as a geometric object, is simply slid onto itself by a motion parallel to the first axis. From this it follows that the cluster C tiles n-space by the vectors of L'.

Next we show that $b_1, \ldots, b_r$ can be chosen so that the first coordinates of $t'_1, \ldots, t'_r$ are rational. Let a_{i1} be the first coordinate of t_i, $1 \leq i \leq r$. Indeed, choose b_i to be any number of the form $-a_{i1} + q_i$, $q_i \in Q$.

The same argument can then be repeated for the second coordinate, and so on, through the nth coordinate. Since the alteration each time affects only one coordinate, the sequence of n steps replaces L by a lattice, with only rational coordinates, which still provides a tiling by C. □

There is a wide range in which to choose the b_i in the preceding proof; b_i could be any number in $-a_{i1} + Q$. In particular, we could choose b_i such that $b_i + a_{i1}$ is not an integer. This proves Theorem 2. The proof of Theorem 3, which concerns the absence of twin cubes, is more involved. For convenience, we restate the theorem.

Theorem 3. *If there is a lattice tiling by a cube without twins, then there is a Q-lattice tiling by a cube without twins.*

Proof. (In this case the cluster C has just one cube.) We wish to show that if there are no twins in the tiling by L, then we can choose the b_i, $1 \leq i \leq r$, as defined in the proof of Theorem 1, to avoid twins in L' and still have $b_i + a_{i1}$ rational.

Since the lattice L' is homogeneous, if it has twins it has twins in which one of the cubes is at the origin. Since we move cubes parallel to the first axis to produce L', we do not introduce twins that share a face perpendicular to that axis. We must show how to avoid the creation of twins perpendicular to any of the other $n - 1$ axes.

Consider, for instance, the second axis. In this case we wish to avoid the presence of the vector $(0, 1, 0, \ldots, 0)$ in L'. Since we alter vectors in L only by a change in the first coordinate, $(0, 1, 0, \ldots, 0)$ can appear in L' only if a vector of the form $(z_2, 1, 0, \ldots, 0)$ is in

L. (The subscript 2 reminds us that we are concerned with twins forming along a surface perpendicular to the second axis.)

Now

$$(z_2, 1, 0, \ldots, 0) = x_{21}t_1 + \cdots + x_{2n}t_n,$$

where at least one of the x_{2i}, $1 \leq i \leq r$, is not 0 since the vector $(z_2, 1, 0, \ldots, 0)$ is moved in our construction. Thus

$$z_2 = x_{21}a_{11} + \cdots + x_{2n}a_{n1},$$

where at least one of the x_{2i}, $1 \leq i \leq r$, is not 0. We must choose $b_1, \ldots, b_r$ so that b_i is in $-a_{i1} + Q$ and

$$z_2 + x_{21}b_1 + \cdots + x_{2r}b_r \neq 0.$$

But before we make such a choice, we must keep in mind that there may be other vectors in L of the form $(z, 1, 0, \ldots, 0)$. Consider another such vector in L, say $(y, 1, 0, \ldots, 0)$. The vector $(y - z_2, 0, \ldots, 0)$, being the difference of vectors in L, is also in L. In fact, it is in K since it is in the same cylinder as $(0, \ldots, 0)$. Thus $y - z_2$ is an integer and therefore $(z_2, 1, 0, \ldots, 0)$ differs from $(y, 1, 0, \ldots, 0)$ by an element of M. Consequently their components in T are identical. Moreover $y = z_2 + u$, where u is an integer. The first coordinate of the translation of $(y, 1, 0, \ldots, 0)$ is then

$$z_2 + u + x_{21}b_1 + \cdots + x_{2r}b_r, \tag{1}$$

a fact that will be used in a moment. In order that (1) is not 0, we must choose the b_i's so that none of the numbers

$$x_{21}b_1 + \cdots + x_{2r}b_r$$

is in the set $-z_2 + Z$.

Before we show that such a choice is possible, recall that twins may be formed that share a face perpendicular to the third axis, and so on. To avoid such twins forming with a face shared perpendicular to the ith axis we must also demand that the choice of $b_1, \ldots, b_r$ does not cause a number of the form

$$x_{i1}b_1 + \cdots + x_{ir}b_r$$

to lie in the set of the form $-z_i + Z$, $2 \leq i \leq n$. (The numbers $x_{i1}, \ldots, x_{ir}$ are analogous to the numbers $x_{21}, \ldots, x_{2r}$.)

Consider again the second axis. Let u be an integer. The set of vectors $(b_1, \ldots, b_r)$ such that

$$x_{21}b_1 + \cdots + x_{2r}b_r = -z_2 + u$$

is an $(r-1)$-dimensional hyperplane in R^r. As u varies through Z, we obtain a set of parallel hyperplanes in R^r at the fixed distance

$$\frac{1}{\sqrt{x_{21}^2 + \cdots + x_{2r}^2}}$$

from each other. As long we choose $(b_1, \ldots, b_r)$ not on any of these hyperplanes, we avoid forming twins that share a face perpendicular to the second axis. Similar reasoning holds for each of the other axes of interest.

To be sure that L' has no twins we choose $(b_1, \ldots, b_r)$ not to lie in any of the $n-1$ families of hyperplanes just mentioned.

That this is possible is a consequence of the fact that b_i can be chosen anywhere in $-a_{i1} + Q$, while consecutive hyperplanes in the $n-1$ families are at a fixed distance from each other. This construction then gives us a lattice L' without twins and such that the first coordinate of each of its vectors is rational.

Then we repeat the process, starting with L', to make all the second coordinates rational. After n successive applications of the same process we obtain a rational lattice without twins. $\qquad\square$

Exercise 6. Write out a careful proof that the union of the $n-1$ families of hyperplanes in the preceding proof is not R^r.

Exercise 7. The proof of Theorem 3 could be simplified by using the following theorem: *R^n is not a denumerable union of subsets of the form $v + W$, where $v \in R^n$ and W is a subspace of dimension at most $n-1$.* Prove this theorem. (Suggestion: start with the case $n = 2$.)

(We should mention that the theorem in Exercise 7 is a special case of a topological theorem: *Let $S_1, S_2, \ldots, S_k, \ldots$ be a denumer-*

able family of closed subsets of R^n. Assume that no set S_k contains a ball. Then the union of the family contains no ball. For a proof see [4, p. 87].)

The proof of Theorem 4, which asserts that if there is a tiling by C then there is a Z-tiling by C, is much easier. First translate the cylindrical equivalence classes so that their vectors all have an integer first coordinate. Starting with this tiling, carry out the same procedure for the second coordinate. After n successive applications of this procedure we obtain a Z-tiling by C.

2. Clusters with a prime number of cubes

We now turn to the theorem concerning clusters with a prime number of cubes.

In addition to Theorem 2, we will need a way to tell whether a Q-lattice tiling is a Z-lattice tiling. We first develop such a method.

Consider a cluster C in R^n, which consists of cubes described by the vectors $c_1, \ldots, c_q$. We also use the same symbol, C, to denote the set of vectors $c_1, c_2, \ldots, c_q$. We may assume that $c_1 = (0, \ldots, 0)$. Let L be the lattice of translating vectors of a Q-lattice tiling of R^n by C.

Since L is a rational lattice there are positive integers $r_1, \ldots, r_n$ such that L is contained in the lattice L' generated by

$$e_1' = (1/r_1)e_1, \ldots, e_n' = (1/r_n)e_n.$$

Moreover, we choose the r_i's to be the smallest positive integers so that $L \subset L'$. If $r_i = 1$, then $e_i' = e_i$ and the ith coordinate of every vector in L is an integer.

That translates of C by the vectors in L tile R^n is equivalent to the assertion that each vector $l' \in L'$ is uniquely expressible in the form

$$l' = l + c_i + x_1 e_1' + \cdots + x_n e_n',$$

where $l \in L$, $c_i \in C$, and x_i is an integer $0 \le x_i \le r_i - 1, 1 \le i \le n$. See Figure 4, which shows a cluster composed of two squares, where r_1 is 1 and r_2 is 3.

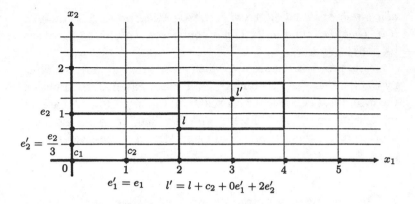

FIGURE 4

Let $D_i = \{0, e'_i, \ldots, (r_i - 1)e'_i\}$, $1 \leq i \leq n$. Then we have the direct sum decomposition

$$L' = L + C + D_1 + \cdots + D_n.$$

This factorization of L' induces a factorization of the quotient group, $G = L'/L$,

$$G = C^* + D_1^* + \cdots + D_n^*, \tag{2}$$

where $C^* = (C + L)/L$ and $D_i^* = (D_i + L)/L$.

Exercise 8. Let the abelian group H be factored as $H = A + B + C$, where A is a subgroup of H and B and C are subsets of H. Prove that this induces a factorization of the group $G = H/A$, namely $G = (B + A)/A + (C + A)/A$.

For convenience we rewrite (2) in multiplicative notation. Letting $a_i = e'_i + L$, we have

$$D_i^* = \{e, a_i, a_i^2, \ldots, a_i^{r_i-1}\}, \qquad 1 \leq i \leq n.$$

Then (2) reads

$$G = C^* D_1^* \cdots D_n^*, \tag{3}$$

where the D_i^* are cyclic subsets of G or just $\{e\}$, where e is the identity element of G.

Now we are ready to develop the test created by Szabó [8] for telling whether a Q-lattice tiling by C is a Z-tiling.

Lemma 2. *Let C be a cluster in R^n that contains the $n+1$ vectors $(0,\ldots,0)$, $e_1,\ldots,e_n$. If C^* in (2) is a subgroup of $G = L'/L$, then $L \subset Z^n$, that is, the Q-lattice tiling by translates of C by the vectors in L is a Z-lattice tiling.*

Proof. Each vector $l \in L$ can be written uniquely in the form

$$l = z_1(1/r_1)e_1 + \cdots + z_n(1/r_n)e_n,$$

where the z_i's are integers. We wish to prove that r_i divides z_i.

In any case, $z_i = u_i r_i + v_i$, where u_i and v_i are integers and $0 \le v_i \le r_i - 1$. We wish to show that $v_i = 0$.

We have

$$l = u_1 e_1 + \cdots + u_n e_n + (v_1/r_1)e_1 + \cdots + (v_n/r_n)e_n. \quad (4)$$

Let $e_i' = (1/r_i)e_i$ and let f be the natural homomorphism $f: L' \to L'/L$. Denote $f(e_i')$ by a_i. Applying f to (4) yields this equation in G:

$$e = (a_1^{r_1})^{u_1} \cdots (a_n^{r_n})^{u_n} a_1^{v_1} \cdots a_n^{v_n}. \quad (5)$$

Since C contains the cubes whose centers are $(0,\ldots,0)$, $e_1,\ldots,e_n$, C^* contains the (distinct) elements $e, a_1^{r_1}, \ldots, a_n^{r_n}$. Therefore, since C^* is assumed to be a group, C^* contains the element

$$(a_1^{r_1})^{u_1} \cdots (a_n^{r_n})^{u_n},$$

which appears in (5). In view of the factorization $G = C^* D_1^* \cdots D_n^*$ and the fact that $e \in C^*$, we conclude from (5) that $a_i^{v_i} = e$. (Otherwise there would be two factorizations of e, the other being $e \cdots e$.) Thus $v_i = 0$ and the tiling is a Z-tiling. $\qquad \square$

We are now ready to prove Theorem 5 in which the cluster has a prime number of cubes.

Theorem 5. *If cluster C has a prime number of cubes and contains the cubes corresponding to $(0, \ldots, 0)$, $e_1, \ldots, e_n$, then any lattice tiling by C is a Z-tiling.*

Proof. If there is a lattice tiling by C that is not a Z-tiling, then, by Theorem 2, there is a Q-lattice tiling by C that is not a Z-tiling. Therefore we may restrict our attention to Q-lattice tilings by C.

Let L, L', G, C^*, D_i^* denote the same sets as in the preceding proof. Delete any D_i^* that consists only of the identity element, $D_i^* = \{e\}$. Relabeling the remaining cyclic sets $D_1^*, \ldots, D_m^*$, $m \leq n$, we have the factorization

$$G = C^* D_1^* \cdots D_m^*. \tag{6}$$

Since $D_i^* = \{e, a_i, a_i^2, \ldots, a_i^{r_i - 1}\}$ and $a_i^{r_i}$ and e are distinct elements of C^*, D_i^* is not a subgroup of G. As we saw in Chapter 1, D_i^* can be expressed as the product of cyclic sets of prime orders that are not subgroups. Rewriting (6) in terms of these sets we now have

$$G = C^* B_1 \cdots B_k, \tag{7}$$

where each factor has prime order and none of the cyclic sets B_i is a subgroup. Rédei's theorem implies that C^* is a subgroup. Lemma 2 then tells us that the tiling has only integer coordinates.   □

Incidentally, in the next chapter we will exhibit clusters C of arbitrarily high dimension n that contain $(0, \ldots, 0)$, $e_1, \ldots, e_n$, have a prime number of cubes, and lattice tile R^n. This assures us that the theorem we just obtained is far from vacuous.

3. Tiling a box by congruent bricks

In the next chapter we return to the study of tiling n-space by translates of some particular clusters. But now we will illustrate some of the many results concerning tiling some bounded region by congruent (not necessarily parallel) copies of a given cluster, by considering the following question that was treated by de Bruijn [1].

For which positive integers a, b, and n can we tile the $a \times b$ rectangle by congruent copies of the $1 \times n$ "brick"? By counting squares we see that if the brick does tile, then n divides ab. Also, if n divides at least one of the integers a and b, then the translates of the brick tile the rectangle. The condition "n divides ab" is not sufficient to assure the existence of a tiling: For instance, the 1×4 brick does not tile the 2×2 rectangle.

We will analyze the problem of determining when the $1 \times n$ brick tiles an $a \times b$ rectangle first by a counting argument. Then we will express the argument algebraically.

To illustrate the idea consider whether the 1×4 brick tiles the 6×10 rectangle. Label the 60 squares 0, 1, 2, or 3, labeling the square (i, j) with the remainder of $i + j$ modulo 4, as in Figure 5. No matter where the brick is placed it covers each of the four symbols the same number of times, namely once. So if there is a tiling, each of the symbols 0, 1, 2, 3 must appear the same number of times in the 6×10 rectangle. However, as may be checked, the four symbols do not appear equally often. Therefore the 1×4 brick does not tile the 6×10 rectangle. This argument could be used to prove the following

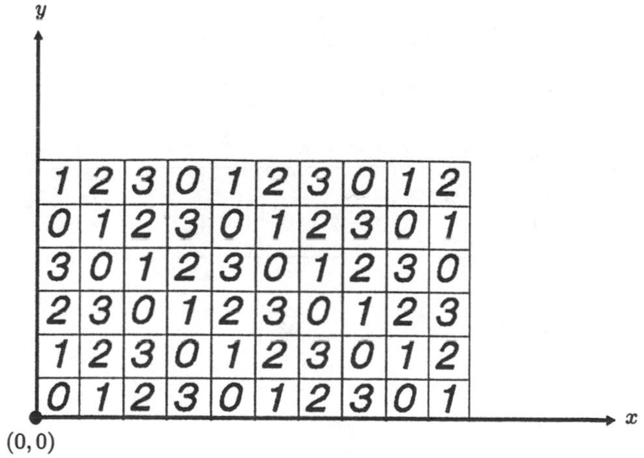

FIGURE 5

theorem. However, it will be easier to prove it algebraically, with polynomials and complex numbers doing the bookkeeping for us. These tools will also come in handy when proving Rédei's theorem in Chapter 7.

Theorem 6. *Let n, a, and b be positive integers. A necessary and sufficient condition that the $1 \times n$ brick tiles the $a \times b$ rectangle is that n divides at least one of a and b.*

Proof. If n divides a or b, then it is possible to tile the $a \times b$ rectangle with parallel bricks. Let us show the converse, that if the $1 \times n$ brick tiles the $a \times b$ rectangle, then n must divide a or b.

Introduce a coordinate system whose axes lie on two of the edges of the $a \times b$ rectangle such that the coordinates of the four vertices of the rectangle are $(0,0)$, $(a,0)$, $(0,b)$, and (a,b). Record each of the ab unit squares that make up the rectangle by the coordinates of its lower left corner, as in Figure 6. For each such corner (i,j), $0 \le i \le a-1$, $0 \le j \le b-1$, introduce the monomial $x^i y^j$. Then record the ab squares in the rectangle by the polynomial

$$\sum_{i=0}^{a-1} \sum_{j=0}^{b-1} x^i y^j$$

which equals $(1 + x + x^2 + \cdots + x^{a-1})(1 + y + y^2 + \cdots + y^{b-1})$.

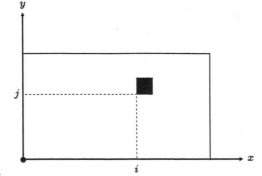

FIGURE 6

When the brick is placed with its long side parallel to the x axis its n cells are recorded by a polynomial of the form

$$x^i y^j + x^{i+1} y^j + x^{i+2} y^j + \cdots + x^{i+n-1} y^j,$$

which equals

$$x^i y^j (1 + x + x^2 + \cdots + x^{n-1}).$$

Similarly, when placed parallel to the y axis it is recorded by a polynomial of the form

$$x^i y^j (1 + y + y^2 + \cdots + y^{n-1}).$$

Therefore if the brick tiles the $a \times b$ rectangle, then there are polynomials $P(x, y)$ and $Q(x, y)$ such that

$$\begin{aligned}
P(x, y)(1 + x + \cdots + x^{n-1}) &+ Q(x, y)(1 + y + \cdots + y^{n-1}) \\
&= (1 + x + \cdots + x^{a-1})(1 + y + \cdots + y^{b-1})
\end{aligned} \tag{8}.$$

Now replace x and y throughout (8) by a primitive nth root of unity, say $\omega = e^{2\pi i/n}$. Since

$$1 + \omega + \omega^2 + \cdots + \omega^{n-1} = \frac{\omega^n - 1}{\omega - 1} = 0,$$

we have

$$P(\omega, \omega)(0) + Q(\omega, \omega)(0) = \left(\frac{\omega^a - 1}{\omega - 1} \right) \left(\frac{\omega^b - 1}{\omega - 1} \right).$$

Therefore at least one of the numbers $\omega^a - 1$ or $\omega^b - 1$ is 0. In the first case n divides a; in the second, n divides b. This concludes the proof. $\qquad\square$

Exercise 9.   When does a $1 \times 1 \times n$ brick tile an $a \times b \times c$ box, where n, a, b, and c are integers?

Exercise 10.   (Extracted from [5]) Let m, n, a, and b be positive integers. Show that if the $m \times n$ rectangle tiles the $a \times b$ rectangle then

(a) each side of the $a \times b$ rectangle can be expressed in the form $xm + yn$, for nonnegative integers x and y;

(b) m divides a or b,

(c) n divides a or b.

Exercise 11. Establish the converse of Exercise 12, in case

(a) m divides a and n divides b (or m divides b and n divides a).

(b) m and n both divide a (or both divide b).

Exercises 10 and 11 answer the question, "When does an $m \times n$ rectangle tile an $a \times b$ rectangle?" The analogous question in higher dimensions has not been settled.

Problem 1. When does an $m_1 \times m_2 \times \cdots \times m_n$ brick tile an $a_1 \times a_2 \times \cdots \times a_n$ n-dimensional box?

Exercise 12. Show that if the $m_1 \times m_2 \times \cdots \times m_n$ brick tiles the $a_1 \times a_2 \times \cdots \times a_n$ box, then each m_i divides at least one of the a_i.

The problem just mentioned is completely solved in the case of a special type of brick, called "harmonic." Let $m_1 \leq m_2 \leq \cdots \leq m_n$ be the dimensions of an n-dimensional brick. If m_i divides m_{i+1} for $1 \leq i \leq n - 1$, we call the brick *harmonic*. An $a_1 \times a_2 \times \cdots \times a_n$ box is called a *multiple* of the $m_1 \times m_2 \times \cdots \times m_n$ brick if there is a permutation ϕ of the indices $1, 2, \ldots, n$ such that m_i divides $a_{\phi(i)}$.

Theorem 7. *If a harmonic brick tiles a box, the box is a multiple of the brick.*

Proof. Let the dimensions of the brick be $m_1, m_2, \ldots, m_n$, where m_i divides m_{i+1}, $1 \leq i \leq n - 1$. Using the technique in the proof of Theorem 6, we see that m_n divides at least one of the dimensions of the box, which we label a_n.

Consider an $(n - 1)$-dimensional face of the box, of dimensions $a_1 \times a_2 \times \cdots \times a_{n-1}$. This face is entirely filled with $(n - 1)$-dimensional bricks of the form $m_1 \times m_2 \times \cdots \times m_{n-1}, \ldots, m_2 \times m_3 \times \cdots \times m_n$. Each of these n types of $(n - 1)$-dimensional bricks is a multiple of the $m_1 \times m_2 \times \cdots \times m_{n-1}$ brick, since the n-dimensional brick is harmonic. By induction, the $a_1 \times a_2 \times \cdots \times a_{n-1}$ box is a multiple of the $m_1 \times m_2 \times \cdots \times m_{n-1}$ brick. Since, in addition, m_n divides a_n, the theorem is proved. $\qquad\square$

Exercise 13. Show that if the $m_1 \times m_2$ brick is not harmonic there is a rectangle that it tiles that is not a multiple of the brick. (Hint: Consider the $(m_1 + m_2) \times m_1 m_2$ rectangle. Try it first for $m_1 = 2$, $m_2 = 3$.)

Exercise 14. What boxes does the $1 \times 2 \times 4$ brick tile?

Exercise 15.

(a) One square is deleted from an 8×8 chessboard. Assume that the remaining 63 squares can be tiled by the 1×3 brick. Using complex numbers, show that the deleted square must be one of the four squares that lie on the two diagonals and share a vertex with one of the four squares at the center of the board. (This proof is due to Mackinnon [6]. A proof using a coloring argument is to be found in Golomb [3].)

(b) Show that if the deleted square is one of the four described in (a) then the 1×3 brick does tile the remaining 63 squares.

As the problem of tiling a box with congruent bricks suggests, deciding whether some bounded or unbounded region can be tiled by congruent copies of some given figures can easily be quite difficult. As another example, consider the triangular array in Figure 7, which is part of the tiling of the plane by regular hexagons.

Denote such a triangle with n hexagons on a side, T_n. Conway and Lagarias [2] considered the question "Which T_n can be tiled by congruent copies of T_2?" Clearly, 3 must divide $n(n + 1)/2$, the total number of hexagons in T_n. So $n \equiv 0$ or 2 (mod 3). Using the free group on two generators they obtained a much stronger result: $n \equiv 0, 2, 9$, or 11(mod 12). Using the same technique, they also proved that T_n can never be tiled by copies of the set of three

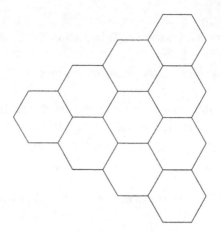

FIGURE 7

hexagons in a line (the middle one being adjacent to the other two). Conway [9] later showed that for $2 < m < n$, T_m does not tile T_n.

Exercise 16. For which integers n can T_n be tiled by copies of a set of two adjacent hexagons?

Deciding whether congruent copies of a given set of clusters tile n-space can also be difficult. In fact, Berger in 1966 and Robinson in 1971 proved that there is no general algorithm for settling the question. In the next chapter we will consider tilings by two particular families of clusters in n-space. Though we construct tilings for many of these clusters, the general question of which ones tile is far from settled.

References

1. N. G. de Bruijn, Filling boxes with bricks, *Amer. Math. Monthly* **76** (1969), 37–40.

2. J. H. Conway and J. C. Lagarias, Tiling with polyominoes and combinatorial group theory, *J. Comb. Theory*, Series A **53** (1990), 183–208.

3. S. Golomb, *Polyominoes*, Princeton Univ. Press, 1994.

4. J. C. Hocking and G. S. Young, *Topology*, Addison-Wesley, Reading, 1961.

5. D. A. Klarner, Packing a rectangle with congruent N-ominoes, *J. Comb. Theory* **7** (1969), 107–115.

6. N. Mackinnon, An algebraic tiling proof, *Math. Gazette* **465** (1989), 210–211.

7. T. Schmidt, Über die Zerlegung des n-dimensionalen Raumes in gitterförmig angeordnete Würfeln, *Schr. math. Semin. u. Inst. angew. Math. Univ. Berlin* **1** (1933), 186–212.

8. S. Szabó, On mosaics consisting of multidimensional crosses, *Acta Math. Acad. Sci. Hung.* **38** (1981), 191–203.

9. W. P. Thurston, Conway's tiling groups, *Amer. Math. Monthly* **97** (1990), 757–773.

Chapter 3
Tiling by the Semicross and Cross

When we hear the expression "convex body" we probably visualize such famous sets as the ball, the cube, or the tetrahedron. However, the expression "non-convex body" triggers no specific image, just the general sense of an object with dents. In this chapter and the next we explore two families of non-convex sets that may well be viewed as the prototypes of non-convex bodies.

These two particular families of clusters in n-space have drawn the attention of mathematicians for several reasons. First, their tiling, packing, and covering properties can be analyzed with the aid of existing algebraic and combinatorial tools. Second, they raise many new questions, even about structures as simple as finite cyclic groups. Third, they are a convenient source of examples and counterexamples for questions concerning bodies that are not convex. Finally, they also appear naturally in such a real-world application as coding theory. In this chapter we define these two families and examine the way they tile n-space. In the next chapter we look at their packings and coverings. At the end of this chapter we sketch their history.

1. Definitions

In this chapter we will restrict our attention to Z-tilings. Recall that if a cluster tiles R^n then it tiles Z^n (Theorem 4 in Chapter 2). However a cluster may lattice tile R^n but not lattice tile Z^n, as was shown in Chapter 2 by the cluster consisting of two squares separated by a square. Throughout, the point $(c_1, \ldots, c_n) \in Z^n$ will represent the cube $(c_1, \ldots, c_n) + Q$, where Q is the cube $\{(x_1, \ldots, x_n) : 0 \le x_i \le 1\}$. In short, we will use Z^n to represent R^n.

For a positive integer k consider the $kn + 1$ points

$$(0, 0, \ldots, 0), \ (i, 0, \ldots, 0), \ (0, i, \ldots, 0), \ldots, (0, 0, \ldots, i),$$

$1 \le i \le k$. Any translate of this set by an element of Z^n is called a (k, n)-*semicross*. Similarly, any translate by elements of Z^n of the set of $2kn + 1$ points

$$(0, 0, \ldots, 0), \ (i, 0, \ldots, 0), \ (0, i, \ldots, 0), \ldots, (0, 0, \ldots, i),$$

$1 \le |i| \le k$, is called a (k, n)-*cross*. Figure 1 shows a $(3, 2)$-semicross and a $(2, 3)$-cross.

Though these clusters are not convex, they are at least "star bodies." A *star body* is a subset of R^n that contains a point P such that for every point Q in it, the entire chord PQ lies in the body, as shown in Figure 2.

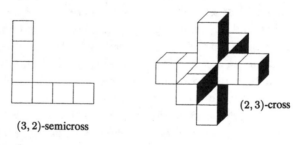

(3, 2)-semicross   (2, 3)-cross

FIGURE 1

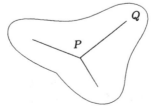

FIGURE 2

The cross and semicross are clusters and so we may apply the definitions of Chapter 2.

We therefore may speak of a tiling or lattice tiling by the (k, n)-semicross or say that the (k, n)-semicross tiles or lattice tiles Z^n. Similar statements hold for the (k, n)-cross.

We immediately face a basic question.

Problem 1.   For which positive integers k and n does the (k, n)-semicross or the (k, n)-cross tile Z^n? Lattice tile Z^n?

Even the lattice tiling part of this problem is far from being resolved. To suggest the complexity of the problem, we mention that the $(k, 10)$-semicross lattice tiles Z^{10} only for $k = 1$, but that the $(k, 12)$-semicross lattice tiles Z^{12} for $k = 1, 2, 3$, and 10.

This chapter presents some of the results motivated by Problem 1.

A $(k, 1)$-semicross is an interval of $k + 1$ points, which tiles Z^1, and only as a lattice. A $(k, 2)$-semicross resembles the letter L and tiles Z^2, again only as a lattice, as shown in Figure 3.

Exercise 1.   Verify the last statement.

A $(k, 1)$-cross is an interval of $2k + 1$ points, which tiles Z^1, and only as a lattice. The $(1, 2)$-cross tiles Z^2, again only as a lattice.

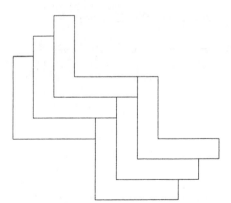

FIGURE 3

Exercise 2.   Verify the last statement.

Exercise 3.   Show that for $k \geq 2$, the $(k, 2)$-cross does not tile Z^2.

It is not hard to show for $n \geq 2$ that if the arm of a cross is too long, then the cross cannot tile Z^n. That is the content of the following theorem [10].

Theorem 1.   *If $n \geq 2$ and $k \geq 2n - 1$, the (k, n)-cross does not tile Z^n.*

Proof.   Suppose that $n \geq 2$ and that the (k, n)-cross tiles Z^n. Consider any translate in Z^n of the $(k + 1)^2$ points in the square array

$$\{(i, j, 0, \dots, 0) : n - 2 \ 0\text{'s}; \ 0 \leq i, j \leq k\}.$$

Figure 4 illustrates the case $k = 3$.

Call such a translate a "tray." Two (k, n)-crosses whose centers lie in a tray overlap. Consider a tiling of Z^n by trays. Since each tray contains at most one center of a cross in the alleged tiling, the density of centers is at most $1/(k + 1)^2$. However, the density of centers is $1/(2kn + 1)$. We have therefore

$$2kn + 1 \geq (k + 1)^2$$

or $k \leq 2n - 2$. This proves the theorem.   □

FIGURE 4

As we will see, the $(2,3)$-cross lattice tiles Z^3 and for an infinite number of even dimensions n, the $((n-2)/2, n)$-cross lattice tiles Z^n. These facts suggest the following problem.

Problem 2.   If the (k,n)-cross tiles Z^n, $n \geq 3$, is k less than n?

For lattice tilings by the (k,n)-cross, $n \geq 3$, it is known that $k \leq n-1$, and we present a proof in the next section. It may not be the best possible result, and we raise the following problem.

Problem 3.   If the (k,n)-cross lattice tiles Z^n, $n \geq 4$, is k less than $n/2$? If it tiles Z^n?

The situation for the semicross is quite different. In this case we do not know whether there is a bound on the length of the arm of a semicross that tiles Z^n, $n \geq 4$. (For $n = 3$ the bound is 1.) On the other hand it is known that for $n \geq 3$ in the case of lattice tilings of Z^n by the (k,n)-semicross, we must have $k \leq n-2$, and this is the best possible general result, though for specific n the bound can be much lower. This is proved in the next section.

Problem 4.   For fixed $n \geq 4$ is there an upper bound on the set of integers k such that the (k,n)-semicross tiles Z^n? (For $n = 1$ or $n = 2$ there is no such bound; for $n = 3$, the bound is 1.)

Assume that the (k,n)-semicross lattice tiles Z^n through translations by the vectors in the subgroup H of Z^n. This condition is equivalent to the fact that the $kn + 1$ elements of the semicross whose corner is at the origin represent the distinct cosets of H. Let $e_j = (0, \ldots, 0, 1, 0, \ldots, 0)$, where the 1 is in the jth place. Let $G = Z^n/H$ and let $f : Z^n \to G$ be the natural homomorphism. Let $s_j = f(e_j)$. Then the elements $\{is_j : 1 \leq i \leq k, 1 \leq j \leq n\}$, together with $0 \in G$, represent each element of G exactly once. In a sense, we may view G schematically as a semicross, as suggested in Figure 5, which corresponds to the case $k = 3$ and $n = 5$.

Conversely, let G be an abelian group of order $kn+1$. Assume that there are n elements, $s_1, s_2, \ldots, s_n$, in G such that each element of $G \setminus \{0\}$ is uniquely expressible in the form is_j, $1 \leq i \leq k$,

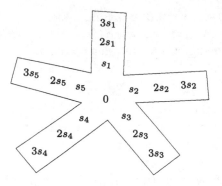

FIGURE 5

$1 \leq j \leq n$. ($G \setminus \{0\}$ denotes the set of nonzero elements of G.) Define $f : Z^n \rightarrow G$ by setting $f(e_j) = s_j$. Then translates of a semicross by the kernel of f lattice tile Z^n.

Exercise 4. Verify the last statement.

If G is an abelian group of order $kn + 1$ in which we can find a set of n elements with the property described, we say that $\{1, 2, \ldots, k\}$ *splits* G, with *splitting set* $\{s_1, s_2, \ldots, s_n\}$. Extensive computations, together with some theorems, suggest that splittings are quite rare.

Similar reasoning shows that the (k, n)-cross lattice tiles Z^n if and only if there is an abelian group G of order $2kn + 1$ in which there are n elements $s_1, s_2, \ldots, s_n$, such that each element in $G \setminus \{0\}$ is uniquely expressible in the form is_j, $1 \leq |i| \leq k$, $1 \leq j \leq n$. If such a set exists we say that $\{\pm1, \pm2, \ldots, \pm k\}$ *splits* G with *splitting set* $\{s_1, s_2, \ldots, s_n\}$.

For convenience, let $S(k) = \{1, 2, \ldots, k\}$ and $F(k) = \{\pm1, \pm2, \ldots, \pm k\}$. ($S$ stands for "semicross" and F stands for "full cross".) So instead of treating lattice tilings of Z^n by the (k, n)-semicross or (k, n)-cross, we examine splittings by $S(k)$ or $F(k)$ of abelian groups of order $kn + 1$ or $2kn + 1$ respectively. The next section uses this approach.

Exercise 5. Show that if the (k, n)-cross Z-lattice tiles R^n, then the $(k, 2n)$-semicross Z-lattice tiles R^{2n}.

This exercise raises the question, "If the (k, n)-cross tiles Z^n, does the $(k, 2n)$-semicross tile Z^{2n}?", which is answered in the next exercise.

Exercise 6. Let S be a finite set of points in Z^n such that $S \cap (-S)$ is the origin of Z^n. ($-S$ denotes the set $\{-s : s \in S\}$.) Assume that $S \cup (-S)$ tiles Z^n by translates by the set H. Let $T = (S, 0) \cup (0, S)$, where 0 is the origin of R^n. Then T is a subset of Z^{2n}. Prove that translates of T tile Z^{2n}. (Suggestion: Use the set of elements of the form $(x, x - h)$, $x \in Z^n$, $h \in H$, as translating vectors.)

2. Bounds on the lengths of the arms

We are now in position to put upper bounds on the length of the arm, k, in case a (k, n)-semicross or (k, n)-cross lattice tiles Z^n. We treat the cross first since the proof, to be found in **[16]**, is shorter.

Theorem 2. *If $n \geq 2$ and the (k, n)-cross lattice tiles Z^n, then $k \leq n - 1$.*

Proof. It is easy to check that the theorem is true for $n = 2$, so we consider only $n \geq 3$. Assume that $F(k)$ splits the finite abelian group G with splitting set $\{s_1, s_2, \ldots, s_n\}$.

We first show that for each integer i, $2 \leq i \leq n$, there are integers x_i, y_i such that $k + 1 \leq x_i \leq 2n - 1$, $|y_i| \leq k$ and $x_i s_1 + y_i s_i = 0$.

For simplicity, take $i = 2$. Consider the $2n(k + 1)$ elements $a_1 s_1 + a_2 s_2$, $0 \leq a_1 \leq 2n - 1$, $0 \leq a_2 \leq k$. Since $2n(k+1) > 2kn+1$, the order of G, there are distinct couples (b_1, b_2) and (c_1, c_2), $0 \leq b_1, c_1 \leq 2n - 1$, $0 \leq b_2, c_2 \leq k$ such that $b_1 s_1 + b_2 s_2 = c_1 s_1 + c_2 s_2$. It is no loss of generality to assume that $b_1 \geq c_1$. Let $d_1 = b_1 - c_1$ and $d_2 = b_2 - c_2$. Then $d_1 s_1 + d_2 s_2 = 0$, where $(d_1, d_2) \neq (0, 0)$ and $0 \leq d_1 \leq 2n - 1$, $|d_2| \leq k$. If $0 \leq d_1 \leq k$, then $d_1 s_1 = -d_2 s_2$, which violates the fact that s_1 and s_2 are part of a splitting set.

Thus for each integer i, $2 \leq i \leq n$, there is a pair of integers (x_i, y_i) such that $k + 1 \leq x_i \leq 2n - 1$, $|y_i| \leq k$, and $x_i s_1 + y_i s_i = 0$.

Assume that there are distinct integers i and j such that $x_i = x_j$. Since $x_i s_1 + y_i s_i = 0$ and $x_j s_1 + y_j s_j = 0$, it follows that $y_i s_i = y_j s_j$. This violates the splitting condition unless $y_i = 0 = y_j$. It follows that $x_i s_1 = 0$.

Note that the $2k + 1$ elements

$$-ks_1, \ldots, -s_1, 0, s_1, \ldots, ks_1$$

are distinct since s_1 is an element of a splitting set. Thus the order of s_1 in G is at least $2k + 1$. Since $x_i s_1 = 0$, the order of s_1 divides x_i; hence $x_i \geq 2k + 1$. But $x_i \leq 2n - 1$. Therefore $2k + 1 \leq 2n - 1$, and $k \leq n - 1$.

If, on the other hand, the $n - 1$ x_i's are distinct, we must have

$$n - 1 \leq 2n - 1 - (k + 1) + 1,$$

since they all lie in the interval $[k + 1, 2n - 1]$. It follows that $k \leq n$.

Next, using the fact that $k \leq n$, we will show that $k \leq n - 1$. Consider the $(2n - 1)(k + 1)$ elements $b_1 s_1 + b_2 s_i$, $0 \leq b_1 \leq 2n - 2$, $0 \leq b_2 \leq k$. Now

$$(2n-1)(k+1) = 2kn + 2n - k - 1 \geq 2kn + n - 1 \geq 2kn + 2 > 2kn + 1.$$

Reasoning as before, we conclude that for each integer i, $2 \leq i \leq n$ there are integers x_i and y_i such that $k + 1 \leq x_i \leq 2n - 2$, $|y_i| \leq k$, and $x_i s_1 + y_i s_i = 0$.

If all $n - 1$ of the x_i's are distinct, then from the fact that they are in the interval $[k+1, 2n-2]$, we have $n - 1 \leq 2n - 2 - (k+1) + 1$; that is, $k \leq n - 1$.

If there is duplication among the x_i's we argue as before, starting with $y_i s_i = y_j s_j$. This time we conclude that $2k + 1 \leq 2n - 2$, hence $k \leq n - 2$.

This concludes the proof. $\square$

Exercise 7.   Fill in any omitted steps in the preceding proof.

For semicrosses, the first argument for the corresponding bound on the arm length was totally algebraic [13]. The following geometric lemma, due to Hickerson, simplifies the proof.

Lemma 1. *Let n and k be integers, $n \geq 3$, $k \geq n - 1$. Assume that $S(k)$ splits an abelian group G of order $nk+1$. Let s and s' be elements of a splitting set. Then one of these two conditions holds:*

(a) There are integers x and y, $1 \leq x \leq n - 2$, $1 \leq y \leq k$, such that $xs + ys' = 0$.

(b) $s' = (1 - n)s$ and G is cyclic with generator s.

Proof. Define $f: Z \oplus Z \to G$ by $f(i, j) = is + js'$. Let $A = \{(i, j): 0 \leq i \leq n - 2, 0 \leq j \leq k\}$.

If $f: A \to G$ is not one-to-one, then there are integers x and y that sastisfy (a). If it is one-to-one, note that it is also one-to-one on $B = A \cup \{(n-1, 0), (n, 0), \ldots, (k, 0)\}$. Now, $|B| = (n-1)(k+1) + k - (n - 2) = nk + 1$. Thus f, considered only on B, is one-to-one from B onto G. Hence B tiles $Z \oplus Z$ by the vectors in the kernel of f. The $nk + 1$ points in B form an L-shaped set, as indicated in Figure 6, where $k = 5$ and $n = 4$.

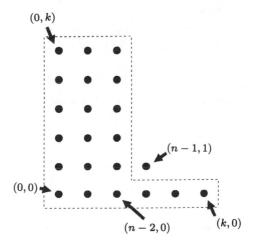

FIGURE 6

Since B tiles $Z \oplus Z$, inspection of Figure 6 shows that $(n-1,1)$ is one of the vectors in the lattice of translating vectors, that is, $(n-1,1)$ lies in the kernel of f. This means that $(n-1)s + s' = 0$.

Since s and s' generate G and $s' = -(n-1)s$, it follows that s generates G, and G is therefore cyclic.   □

With the aid of the lemma, the next theorem follows quickly.

Theorem 3. *Let n and k be integers, $n \geq 3$. If $S(k)$ splits a group G of order $nk + 1$, then $k \leq n - 2$.*

Proof. Suppose $k \geq n-1$. Let $s_1, s_2, \ldots, s_n$ be a splitting set of G, a group of order $nk + 1$. For each index j, $2 \leq j \leq n$, consider the pair of elements s_1 and s_j. Assume that for each such j, condition (a) in Lemma 1 holds, that is, there are x_j and y_j, $1 \leq x_j \leq n-2$, $1 \leq y_j \leq k$, such that $x_j s_1 + y_j s_j = 0$. There would then be $n-1$ values of x_j in the interval $[1, n-2]$. By the pigeon-hole principle, two of these would be equal, say $x_u = x_v$. It follows that $y_u s_u = y_v s_v$, violating the assumption that s_u and s_v are in a splitting set.

Thus there must be an index j such that condition (b) in Lemma 1 holds. That means that G is cyclic, s_1 is a generator of G, and $(1-n)s_1$ is in the splitting set.

The same reasoning, with $(1-n)s_1$ playing the role of s_1, shows that $(1-n)(1-n)s_1$ is also in the splitting set. Since s_1 is a generator of G, and the order of G is larger than $(1-n)^2$, the elements s_1 and $(1-n)^2 s_1$ are distinct. However, since

$$k(1-n)^2 \equiv k + 2 - n \pmod{kn + 1},$$

$$k(1-n)^2 s_1 = (k + 2 - n)s_1,$$

violating the fact that $(1-n)^2 s_1$ and s_1 are in the splitting set.   □

As Exercise 12 will show, the general bound $k \leq n-2$ in Theorem 3 cannot be lowered. However, for particular n, it may be much lower.

3. The search for splittings

Lattice tilings of Z^n by the semicross or cross led us to introduce the notion of splitting an abelian group by certain sets of integers $S(k) = \{1, 2, \ldots, k\}$ and $F(k) = \{\pm 1, \pm 2, \ldots, \pm k\}$. The question,"Which groups do $S(k)$ or $F(k)$ split?," is far from being answered.

For example, $S(k)$ splits $C(k + 1)$ with splitting set $\{1\}$, and splits $C(2k + 1)$ with splitting set $\{-1, 1\}$. ($C(m)$ denotes the cyclic group of order m, which we will usually take to be $\{0, 1, \ldots, m-1\}$, the additive integers modulo m.) However, it is not known whether $S(k)$ always splits some other group. In particular, the answer is not known for $S(195)$ [1], which we will discuss later.

Exercise 8.   Show that if $F(k)$ splits G, then $S(k)$ also splits G.

In order to make general statements about which groups $S(k)$ or $F(k)$ splits, we develop some theorems that relate splittings of groups A and B to splittings of their direct sum $A \oplus B$. And, rather than restricting ourselves to the sets $S(k)$ and $F(k)$, we consider any set of nonzero integers $M = \{m_1, m_2, \ldots, m_k\}$, which we call a *multiplier set*. We say that M *splits* the finite abelian group G if there is a set of elements $s_1, s_2, \ldots, s_n$ in G such that each nonzero element of G is uniquely representable in the form $m_i s_j$, $1 \leq i \leq k$, $1 \leq j \leq n$, and 0 is not representable in that form. It follows that $|G| = kn + 1$. S is called a *splitting set* for G.

Let A, B, and G be groups such that there is an isomorphism α from A into G, a homomorphism β from G onto B, and the kernel of β coincides with the image of α. We may think of A as a subgroup of G and B as the quotient group G/A. (This includes the case when G is a direct sum of A and B.) We record this by the diagram

$$\{0\} \longrightarrow A \xrightarrow{\ \alpha\ } G \xrightarrow{\ \beta\ } B \longrightarrow \{0\},$$

where each arrow denotes a homomorphism. The groups $\{0\}$, A, G, B, $\{0\}$ and the four homomorphisms are said to form an "exact sequence." (See Appendix B for an introduction to exact sequences.)

Exercise 9. Show that if $G = A \oplus B$, then there is an exact sequence

$$\{0\} \longrightarrow A \xrightarrow{\alpha} G \xrightarrow{\beta} B \longrightarrow \{0\}.$$

For any finite abelian group G two questions immediately come into mind: If a multiplier set M splits G, must it split A and B? If M splits A and B, must it split G? The answers to both questions are "no," as we show by examples. However, with an extra assumption, the answers turn to "yes."

First of all, $M = S(5)$ splits $C(6)$ and though there is an exact sequence

$$\{0\} \longrightarrow C(2) \longrightarrow C(6) \longrightarrow C(3) \longrightarrow \{0\},$$

it does not split either $C(2)$ or $C(3)$. So the first question is settled.

Exercise 10. Show that if M splits the cyclic group $C(m)$, then there is a splitting set that contains 1.

To show that the answer to the second question is also "no," consider $M = S(3)$ and the exact sequence

$$\{0\} \longrightarrow C(4) \longrightarrow C(16) \longrightarrow C(4) \longrightarrow \{0\}.$$

Note that M splits the two "outer" groups. (The first $C(4)$ has the elements 0, 4, 8, and 12. The homomorphism from it to $C(16)$ is inclusion. The homomorphism from $C(16)$ to $C(4)$ is the remainder modulo 4.) We show that it does not split $C(16)$. Consider $D = \{1, 3, 5, 7, 9, 11, 13, 15\} \subset C(16)$. The mapping $d \to 3d$ of D consists of two cycles, $(1, 3, 9, 11)$ and $(5, 15, 13, 7)$. There is no loss of generality to assume that if there is a splitting set for $C(16)$, then there is one that contains 1. We call it S. It follows that $3 \notin S$. But, in order that 9 be covered, we must have $9 \in S$. However, $2 \cdot 1 = 2 \cdot 9$, which shows that S is not a splitting set.

Fortunately, there are some relations between splittings of a group, a subgroup, and a quotient group. These relations, developed in [2, 5, 6], are expressed in the following theorems, which greatly simplify the search for splittings.

Theorem 4. *Let M split the finite abelian group G. Assume that there is an exact sequence*

$$\{0\} \longrightarrow A \xrightarrow{\alpha} G \xrightarrow{\beta} B \longrightarrow \{0\}. \tag{1}$$

If each element of M is relatively prime to the order of B, then M splits A.

Proof. Let M split G, with splitting set S. We show that M splits $\alpha(A)$ with splitting set $\alpha(A) \cap S$.

Let $a \in A$. Then $\alpha(a) = ms$, $m \in M$, $s \in S$. Applying β, we have $0 = \beta\alpha(a) = m\beta(s)$. Since $(m, |B|) = 1$, $\beta(s) = 0$. By exactness, $s \in \alpha(A)$. Thus M splits $\alpha(A)$, hence A. $\qquad\square$

If G is a finite abelian group and there is an exact sequence (1), then there is also an exact sequence

$$\{0\} \longrightarrow B \xrightarrow{\bar{\beta}} G \xrightarrow{\bar{\alpha}} A \longrightarrow \{0\},$$

that is, the roles of image and kernel can be switched. (A proof is outlined in Appendix B.) Consequently, the next theorem follows immediately from Theorem 4.

Theorem 5. *Let M split the finite abelian group G. Assume that there is an exact sequence*

$$\{0\} \longrightarrow A \xrightarrow{\alpha} G \xrightarrow{\beta} B \longrightarrow \{0\}.$$

If each element of M is relatively prime to the order of A, then M splits B.

While the proof of Theorem 4 shows how to find the promised splitting set of A, the proof of Theorem 5 does not give a direct way to derive the splitting set of B from the splitting set of G.

Problem 5. Find a direct proof of Theorem 5.

The next theorem shows when splittings of A and B in (1) imply the existence of a splitting of G.

Theorem 6. *Let M split the finite abelian groups A and B and assume there is an exact sequence*

$$\{0\} \longrightarrow A \xrightarrow{\alpha} G \xrightarrow{\beta} B \longrightarrow \{0\}.$$

If each element of M is relatively prime to the order of A or each element of M is relatively prime to the order of B, then M splits G.

Proof. In the first case the proof is constructive. Let $S_A = \{s_1, s_2, \ldots, s_a\}$ be a splitting set for A and $S_B = \{t_1, t_2, \ldots, t_b\}$ be a splitting set for B. Pick $u_j \in G$, $1 \leq j \leq b$ such that $\beta(u_j) = t_j$. Then we assert that $S = \{\{u_1, u_2, \ldots, u_b\} + \alpha(A)\} \cup \alpha(S_A)$ is a splitting set for G.

To see this, consider $g \in G$, $g \neq 0$. If $\beta(g) = 0$, then $g \in \alpha(A)$, and is of the form $m_i \alpha(s_j)$, $s_j \in S_A$. If $\beta(g) \neq 0$, it is of the form $m_i t_j$, for some $t_j \in S_B$. Thus $\beta(g - m_i u_j) = 0$ and $g - m_i u_j = \alpha(a)$ for some element $a \in A$. Since $(m_i, |A|) = 1$, a can be written as $m_i a'$, $a' \in A$. Hence $g = m_i u_j + \alpha(m_i a') = m_i(u_j + \alpha(a'))$. Thus every $g \in G \setminus \{0\}$ has a representation in the form $g = m_i s$, $m_i \in M$, $s \in S$, and it is easy to check that the representation is unique.

The second case follows from the first by reversing the exact sequence. $\square$

Exercise 11. Show that the representation defined in the proof of Theorem 6 is unique.

Theorems 4, 5, and 6 suggest that we distinguish between two types of splittings. A splitting of a group G in which each element m in the multiplier set M is relatively prime to $|G|$ is a *nonsingular* splitting. Splittings that are not nonsingular are *singular*. It is convenient to distinguish among the singular splittings the *purely singular*, splittings in which each prime that divides $|G|$ divides at least one element in the multiplier set.

Theorems 4, 5, and 6 tell us that if there is an exact sequence

$$\{0\} \longrightarrow A \xrightarrow{\alpha} G \xrightarrow{\beta} B \longrightarrow \{0\},$$

where G is a finite abelian group, then M splits G nonsingularly if and only if M splits A and B nonsingularly.

For example, if p is a prime that divides no element of M, then M splits $C(p^2)$ if and only if M splits $C(p)$, since there is an exact sequence

$$\{0\} \longrightarrow C(p) \longrightarrow C(p^2) \longrightarrow C(p) \longrightarrow \{0\}.$$

In view of this, the exact sequence

$$\{0\} \longrightarrow C(p) \longrightarrow C(p^3) \longrightarrow C(p^2) \longrightarrow \{0\},$$

then tells us that M splits $C(p^3)$ if and only if M splits $C(p)$. Induction shows that M splits $C(p^n)$ nonsingularly if and only if it splits $C(p)$.

Exercise 12. Prove that if p is a prime, $k = p - 1$, and $n = p + 1$, then the (k, n)-semicross lattice tiles n-space.

An argument similar to that before Exercise 12 establishes a more general theorem.

Theorem 7. *Let G be a finite abelian group and let $p_1, p_2, \ldots, p_r$ be the prime divisors of the order of G. Then M splits G nonsingularly if and only if M splits $C(p_i)$, for each i, $1 \le i \le r$.*

Exercise 13. Prove Theorem 7.

According to Theorem 7, if each element of M is relatively prime to $|G|$, then to determine whether M splits G, it suffices to determine whether M splits $C(p)$ for each prime p that divides $|G|$.

Exercise 14. What can we say about $|M|$ if M splits a group of order $5^3 7^{10}$ nonsingularly?

In order to examine splittings of a finite abelian group G for any multiplier set M, we express G as the sum of two special subgroups A and B: B is the sum of the Sylow subgroups corresponding to primes that divide no element in M; A is the sum of the Sylow subgroups corresponding to primes that divide at least one element in M. By Theorem 6, if M splits A and B, then it splits G. Theorem 4 is almost the converse, showing that if M splits G it splits A. That the full converse holds, and M also splits B, is a consequence of the following theorem. The proof, which uses a counting argument, depends on the finiteness of the group G. Without the assumption of finiteness, the theorem does not hold. To see this, note that $M = \{-1, 1\}$ splits the infinite cyclic groups Z and $4Z$ but not their quotient group $C(4)$. (Splitting of infinite abelian groups is defined like that of finite groups.)

Theorem 8. *Let G be a finite abelian group and*

$$\{0\} \longrightarrow A \xrightarrow{\alpha} G \xrightarrow{\beta} B \longrightarrow \{0\}$$

an exact sequence. If M splits G and B, it splits A. (By the interchangeability of A and B, it follows that if M splits G and A, it splits B.)

We omit the rather involved counting argument due to Hickerson [6], which, in a nutshell, is this: Let G be a finite abelian group and H a subgroup of G. Assuming that M splits G with splitting set S and also splits the quotient group G/H, one shows that $S \cap H$ is a splitting set for H.

Problem 6. Obtain a direct proof of the second part of Theorem 8.

Exercise 15. (If the proof of Theorem 8 has been read.) Let G be a finite group, not necessarily abelian, and let H be a normal subgroup of G. Show that if M splits both G and G/H, then M splits H. (How would you define the splitting of any group, abelian or not?)

Problem 7. Let G be a finite group, not necessarily abelian, and let H be a normal subgroup of G. If M splits H and G, must M split G/H?

Exercise 16. Let G be an infinite abelian group and H a subgroup of G. Show that if M splits G and G/H, it need not split H. (Hint: Let G be the dyadic rationals modulo 1, $H = \{0, 1/2\}$, and $M = \{1, 2\}$.)

Exercise 17. Show that $S(1)$ splits every finite abelian group. What does this imply about the $(1, n)$-semicross?

Exercise 18. Show that $F(1)$ splits every finite abelian group of odd order. What does this imply about the $(1, n)$-cross?

Exercise 19. Decide whether $S(2)$ splits $C(p)$, $p = 3, 5, 7, 11, 13,$ $17, 19, 23,$ and 29.

Exercise 20. Show that the $(2, 3)$-cross lattice tiles Z^3.

Exercise 21. Does $F(2)$ split $C(5)$, $C(17)$, $C(85)$? What does this say about certain crosses?

Exercise 22. Prove that $S(8)$ does not split $C(105)$.

Exercise 23. Prove that $S(11)$ does not split $C(210)$.

The next section uses some of the theorems proved in this section to examine splittings by $S(k)$ and $F(k)$.

4. Splitting by $S(k)$ and $F(k)$

To find out whether the (k, n)-semicross lattice tiles Z^n, we must find out whether $S(k)$ splits an abelian group of order $kn + 1$. And similarly, to find out whether the (k, n)-cross lattice tiles Z^n we must find out whether $F(k)$ splits an abelian group of order $2kn+1$. In view of the theorems of the preceding section, we face two questions:

What are the purely singular splittings for $S(k)$?

For which primes p does $S(k)$ split $C(p)$?

These questions and the analogous ones for $F(k)$ are far from being answered, and we present some of the results obtained so far. We first obtain a theorem on purely singular splittings by $S(k)$ or $F(k)$.

Let G be a finite abelian group and p be a prime. A *p-group* is a group such that the order of each element in it is a power of p. The p-dimension of G (written "$\dim_p G$") is the number of p-groups in the factorization of G as a direct sum of cyclic groups of prime power orders. This direct sum is called the *p-component* of G. It is the unique p-Sylow subgroup of G.

The next exercise is the key to Lemma 2, which can be found in [6].

Exercise 24. Let p be a prime divisor of the order of the finite abelian group G. Assume that M splits G with splitting set S. Show that for $m \in M$ and $s \in S$, ms is not of the form pg, $g \in G$ if and only if p does not divide m and s is not of the form ph, $h \in G$.

Lemma 2. *Let G be a finite abelian group such that M splits G with splitting set S. Let p be a prime divisor of $|G|$ and $\delta_p(M)$ be the number of elements of M that are divisible by p. Then*

$$p^{\dim_p G} \delta_p(M) < |M|.$$

Proof. Let $pG = \{pg : g \in G\}$ and $B = G \setminus pG$. Then

$$|S \cap B| = \frac{|B|}{|M| - \delta_p(M)} = \frac{|G| - |pG|}{|M| - \delta_p(M)}.$$

A straightforward counting argument shows that

$$|pG| = \frac{|G|}{\left(p^{\dim_p G}\right)}.$$

Thus

$$|G| > |G| - 1 = |M||S| \geq |M||S \cap B|$$
$$= \frac{|M||G|(1 - p^{-\dim_p G})}{|M| - \delta_p(M)},$$

or

$$|M| - \delta_p(M) > |M|(1 - p^{-\dim_p G}),$$

from which the lemma quickly follows.   □

Exercise 25. Verify the claim in the preceding proof that $|pG| = |G|/p^{\dim_p G}$.

Corollary 1. *Let G be a finite abelian group with a splitting $G\backslash\{0\} = MS$. Let p be a prime divisor of $|G|$ such that the p-component of G is not cyclic. Then*

$$\delta_p(M) < |M|/p^2.$$

Corollary 2. *Let $M = S(k)$ or $F(k)$ and MS be a purely singular splitting of the finite abelian group G. Then G is cyclic.*

Proof. When M is $S(k)$ or $F(k)$ and $p \leq k$, it is not hard to show that

$$\delta_p(M) \geq |M|/p^2.$$

Corollary 1 then implies that G is cyclic.   □

Exercise 26. Carry out the details of the proof of Corollary 2.

Corollary 2, together with Theorems 5, 6, 7, and 8, yields the following theorem, again found in [**6**].

Theorem 9. *If M is $S(k)$ or $F(k)$ and splits the finite abelian group G, then M splits the cyclic group of the same order as G, $C(|G|)$.*

Proof. Let P be the set of prime divisors of $|G|$ that divide at least one element of M. Let Q be the set of prime divisors of $|G|$ that divide no element of M.

Let

$$H = \sum_{p \in P} \mathrm{Syl}_p(G) \quad \text{and} \quad K = \sum_{q \in Q} \mathrm{Syl}_q(G).$$

Then $G = H \oplus K$. By Theorem 5, M splits H. Corollary 2 shows that H is cyclic. By Theorem 8, M splits K. Theorem 7 then shows that M splits the cyclic group $C(|K|)$. Since M splits H and $C(|K|)$ we conclude from Theorem 6 that M splits $H \oplus C(|K|)$, which is the cyclic group $C(|G|)$. $\qquad\qquad\square$

Exercise 27. Is the converse of Theorem 9 true? That is, if $S(k)$ or $F(k)$ splits $C(m)$ does it split every abelian group of order m?

Corollary 2 asserts that when $S(k)$ splits G purely singularly, then G is cyclic. What are these purely singular splittings? For instance, if $k + 1$ is composite, then $S(k)$ splits $C(k + 1)$ purely singularly and if $2k + 1$ is composite $S(k)$ splits $C(2k + 1)$ purely singularly. Hickerson has shown, in manuscript, that for $k \leq 3000$, the only purely singular splittings by $S(k)$ are of $C(k + 1)$ or $C(2k + 1)$ as just described.

Problem 8. Find all the purely singular splittings by $S(k)$.

Exercise 28. Show that $S(2)$ does not have any purely singular splittings.

Exercise 29. If the result cited for $k \leq 3000$ holds for all k, for which values of k does $S(k)$ have *no* purely singular splittings?

Exercise 30. Prove that the only purely singular splitting by $S(4)$ is of $C(9)$.

Now consider the nonsingular splittings by $S(k)$. In this case we ask, "Which groups of prime order does $S(k)$ split?" In geometric terms this question is equivalent to "If a (k, n)-semicross has a prime number of cubes, when does it lattice tile Z^n?" In view of Theorem 5 of Chapter 2, this in turn is equivalent to, "If a (k, n)-semicross has a prime number of cubes, when does it lattice tile n-space?"

For instance, if $k+1$ or $2k+1$ is prime, $S(k)$ splits at least one such group. In the first case the splitting set is $\{1\}$ and in the second, $\{-1, 1\}$. In both cases these splitting sets are subgroups of the multiplicative group $C(p)^*$, so $S(k)$ can be viewed as coset representatives of a subgroup of $C(p)^*$. Strictly speaking, $S(k)$, which consists of integers, is not a subset of $C(p)^*$, which consists of the residue classes (mod p) relatively prime to p. However, we may identify $S(k)$ with subsets of $C(p)^*$ without causing any confusion. Indeed we then have a factorization of the group $C(p)^*$, $C(p)^* = S(k)H$, where H is a subgroup of $C(p)^*$, namely, the unique subgroup of order $(p-1)/k$.

These splittings suggest that we first look for cases where $S(k)$ splits $C(p)^*$ by being the coset representatives of the subgroup H of order $(p-1)/k$, the splitting set then being H itself. If there is such a subgroup H, we say that $S(k)$ *coset splits* $C(p)$.

Consider, for instance $S(6)$ and $C(103)$. When the elements of $C(103)^*$ are expressed as powers of the generator 5, we have

$$1 = 5^0,\ 2 = 5^{44},\ 3 = 5^{39},\ 4 = 5^{88},\ 5 = 5^1,\ 6 = 5^{83}.$$

(The exponents, which behave like logarithms, are called indices.) The subgroup H of order $(103-1)/6 = 17$ consists of the numbers 5^{6j}, $0 \le j \le 16$. Since the indices of 1, 2, 3, 4, 5, 6 are incongruent modulo 6, $S(6)$ represents the six cosets of H in $C(103)^*$, and $S(6)$ coset splits $C(103)$. (It follows that the $(6, 17)$-semicross lattice tiles Z^{17}.) The only primes $p < 3584$ such that $S(6)$ coset splits $C(p)$ are

$$7, 13, 103, 487, 547, 823, 967, 1063, \text{ and } 3187.$$

As we will see in a moment, $S(6)$ coset splits $C(p)$ for an infinity of primes p.

Exercise 31. For which primes p does $S(2)$ coset split $C(p)$? Express the answer in terms of the order of 2 modulo p.

We will show that if $S(k)$ coset splits at least one group of prime order, it splits an infinite number of them [10]. In order to do this we need to express the notion of coset splitting in terms of homomorphisms.

Assume that $S(k)$ coset splits $C(p)$, with splitting group H. There is then a natural homomorphism from $C(p)^*$ onto the quotient group $C(p)^*/H$, which is isomorphic to $C(k)$. So there is an epimorphism

$$f : C(p)^* \to C(k), \tag{2}$$

that is one-to-one on the subset $\{1, 2, \ldots, k\}$. (In $C(p)^*$ the operation is multiplication; in $C(k)$ it is addition.)

Exercise 32. Prove that if there is a homomorphism from $C(p)^*$ onto $C(k)$ that is one-to-one on $S(k)$, then $S(k)$ coset splits $C(p)$.

Let ϕ be the restriction to $S(k)$ of the homomorphism (2). This function has two properties:
 (i) ϕ is one-to-one from $S(k)$ onto $C(k)$
 (ii) If x, y, and xy are in $S(k)$, then $\phi(xy) = \phi(x) + \phi(y)$.
  Any function from $S(k)$ to $C(k)$ that satisfies (i) and (ii) we call a *logarithm* on $S(k)$. A logarithm on $S(k)$ is determined by its values on the primes in $S(k)$.
  For example the following table describes a logarithm on $S(7)$.

x	1	2	3	4	5	6	7
$\phi(x)$	0	1	3	2	6	4	5

The assigment $\phi(2) = 1$ determines $\phi(4) = \phi(2) + \phi(2) = 1 + 1 = 2$. The choice $\phi(3) = 3$ then forces $\phi(6) = \phi(2 \cdot 3) = 1 + 3 = 4$. Since 5 and 7 are primes in the range $(7/2, 7]$, the values, $\phi(5)$ and $\phi(7)$ may be assigned freely, as long as ϕ is one-to-one.

Exercise 33. Construct logarithms on $S(k)$ for $k = 13$ and 19.

Exercise 34. Show that there is logarithm on $S(k)$ for $k \leq 24$. Recall that there is if $k + 1$ or $2k + 1$ is prime.

Extensive computer calculations by Forcade and Pollington [1], show that for $k \leq 194$, $S(k)$ has a logarithm, but $S(195)$ does not.

Exercise 35. Prove that if there is a logarithm ϕ on $S(k)$ in which $\phi(2)$ is relatively prime to k, then there is a logarithm ψ on $S(k)$ such that $\psi(2) = 1$.

Since there is no logarithm on $S(195)$ it follows that $S(195)$ does not coset split any group of prime order. This raises a sequence of questions.

Problem 9. Does $S(195)$ split some group of prime order? (This is equivalent to "Does the $(195, n)$-semicross, $n \geq 3$, ever lattice tile Z^n?")

Problem 10. Does the $(195, n)$-semicross, $n \geq 3$, ever lattice tile R^n? (This is equivalent to Problem 9.)

Problem 11. Does the $(195, n)$-semicross, $n \geq 3$, ever tile R^n?

We have shown that if $S(k)$ coset splits $C(p)$ for some p, then there is a logarithm on $S(k)$. If k is odd, the converse holds. However, when k is even, the logarithm must satisfy extra conditions in order for the converse to hold, as shown by Mills [8]. These conditions are consequences of quadratic reciprocity.

Theorem 10. *Let $p_1, \ldots, p_r$ be distinct primes and let $b_1, \ldots, b_r \in C(k)$. There is an infinite number of primes p and homomorphisms $\phi : C(p)^* \to C(k)$ such that $\phi(p_i) = b_i$, for $1 \leq i \leq r$ if and only if one of these conditions holds:*

(1) k *is odd*
(2) $k = 2m$ *where m is odd and*
  (a) *for each $p_i \equiv 1 \pmod 4$ that divides m, b_i is even, and*
  (b) *for each $p_i \equiv 3 \pmod 4$ that divides m, the corresponding b_i's all have the same parity.*
(3) $k = 4m$ *and for each p_i that divides m, b_i is even.*

  Moreover, if there is one such prime p for which there is a homomorphism $\phi : C(p)^* \to C(k)$ with the prescribed values at $p_1, \ldots, p_r$, then there is an infinite number of such primes.

Exercise 36. Show that if $S(k)$ coset splits some group of prime order it splits an infinite number of groups of prime order.

Exercise 37. Using the quadratic reciprocity theorem, show that conditions (2) and (3) are necessary when k is even.

Exercise 38. Prove that the $(32, n)$-semicross tiles Z^n for an infinite number of n.

Exercise 39. Let $a_2, a_3, \ldots, a_k$ be distinct integers greater than or equal to 2. Prove that $\{1, \pm a_2, \pm a_3, \ldots, \pm a_k\}$ splits no abelian group (finite or infinite). (Suggestion: If $s_1, s_2, \ldots$ is an alleged splitting set, consider the representation of $-s_1$.)

A special case of Exercise 39 is the set $\{1, \pm 2, \ldots, \pm k\}$, which almost coincides with $F(k)$. This shows that just a slight alteration of a multiplier set can drastically change the groups it splits.

5. History and applications

The origins of the study of the cross and semicross are simple, though they can be traced back to several independent sources: Ulrich in 1957 [18], Kárteszi in 1966 [7], Stein in 1967 [10], and Golomb and Welch in 1968 [3].

Ulrich constructed single-error correcting codes for alphabets of more than two symbols. He utilized a packing of $\{-1, 1\}$ in $C(10)$, presented the equivalent of a splitting of $C(5) \oplus C(5)$ by $\{-1, 1\}$. However, this paper did not lead to subsequent investigations of crosses or semicrosses in Euclidean space.

Kárteszi asked whether the $(1, 3)$-cross tiles space. This was answered by Freller in 1970; Korchmáros about the same time treated $n > 3$. Molnár [9] in 1971 related the number of Z-lattice tilings of R^n by the $(1, n)$-cross to the number of abelian groups of order $2n + 1$. Medyanik, apparently unaware of Molnár's work, showed in 1977 that the $(1, n)$-cross tiles R^n.

Around 1963 Stein posed the following problem. Consider the standard lattice of unit squares that tile the plane. What is the smallest density of a set S of such squares with the property that every square from the lattice has at least one edge on the border of a square in S? That the answer is $1/5$ follows immediately from the fact that the $(1, 2)$-cross tiles the plane. (Each such cross must contain at least one member of S; hence, the density of S is at least $1/5$. On the other hand, the set of the center squares of the crosses in the tiling serves as a suitable family S.) This initiated his work in (k, n)-crosses and semicrosses, which first appeared in 1967.

Golomb and Welch showed that the $(1, n)$-cross tiles R^n. They thought of the center of a cross as a code word and the other cubes of the cross as words that might be received if there were an error in one coordinate of the code word. A tiling then corresponds to a perfect code.

In 1978, Szabó [14], stimulated by Molnár's work, considered tilings by "lopsided" crosses, where at each facet of the central cube either no cube or one cube is attached. Around that time he read a Russian translation of a paper by Stein and became familiar with the work of Hamaker and Stein [5]; in 1981 he proved that if $2n + 1$ is not a prime, then there is a nonlattice Z-tiling by the $(1, n)$-cross and a Q-lattice tiling that is not a Z-tiling [14].

The semicross and cross have also shown that they deserve to serve as archetypes of starbodies. For instance, Kasimatis in 1984 proved that any lattice covering of R^2 by translates of a $(2, 2)$-cross has density at least $9/7$, which is less dense than the densest covering. This example, which is still in manuscript, is much simpler than the earlier ad hoc example.

In 1985 Szabó [17] modified a $(1, 3)$-cross by adding pyramids at the ends of its six arms to obtain a starbody that tiles R^3 but not by a group of motions, thus providing a simple answer to Hilbert's 18th problem, "If congruent copies of a polyhedron P tile Euclidean space, is there a group of motions such that copies of P under this group of motions tile space?" (See Figure 7. The pyramids are lopsided to prevent rotational symmetry.)

Other examples have been constructed even in the plane, but this is a particularly simple example. It also is an example of a set that tiles by translates, but not by a lattice of translates. Stein in 1972 [11] had used a $(4, 10)$-cross and $(3, 5)$-semicross and finite fields to provide the first examples of this phenomenon.

In this chapter we answered some questions concerning the tiling of n-space by the (k, n)-semicross and (k, n)-cross. However, as is typical in mathematics, the few answers obtained raised many questions. This suggests that our knowledge, measured by the number of answers, grows arithmetically, while our ignorance, measured

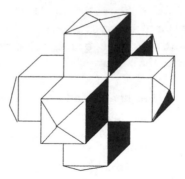

FIGURE 7

by the number of questions, grows geometrically. So, strangely, we are perhaps more ignorant at the end of the chapter than at the beginning.

The tiling problems suggested questions about finite abelian groups. Because of the rich structure of a splitting—each nonzero element of a finite abelian group being represented exactly once—we could use counting arguments to help settle some of the questions.

In the next chapter we will examine packing and covering by the semicross and cross. The structures will not be so rich, and the problems will therefore be more difficult to approach.

References

1. R. W. Forcade and A. D. Pollington, What is special about 195?, Groups, nth power maps and a problem of Graham, *Number Theory*, Richard A. Mollin (Ed.), Walter de Gruyter, New York, 1990.

2. S. Galovich and S. Stein, Splittings of abelian groups by integers, *Aequationes Math.* **22** (1981), 249–267.

3. S. Golomb and L. Welch, Perfect codes in the Lee metric, *University of Southern California, USCEE report 249* (1968), 1–24.

4. W. Hamaker, Factoring groups and tiling space, *Aequationes Math.* **9** (1973), 145–149.

5. W. Hamaker and S. Stein, Splitting groups by integers, *Proc. Amer. Math. Soc.* **46** (1974), 322–324.

6. D. R. Hickerson, Splittings of finite abelian groups, *Pacific J. Math.* **107** (1983), 141–171.

7. F. Kárteszi, *Szemléletes geometria*, Gondolat, Budapest, 1966.

8. W. H. Mills, Characters with preassigned values, *Canad. J. Math.* **15** (1963), 169–171.

9. E. Molnár, Sui mosaici dello spazio di dimensione n, *Atti della Acadamia Nazionale dei Lincei, Rend. Sc. Fis. Mat. e Nat.* **51** (1971), 177–185.

10. S. K. Stein, Factoring by subsets, *Pacific J. Math.* **22** (1967), 523–541.

11. ——, A symmetric star body that tiles but not as a lattice, *Proc. Amer. Math. Soc.* **36** (1972), 543–548.

12. ——, Tiling, packing and covering by clusters, *Rocky Mountain J. Math.* **16** (1986), 277–321.

13. ——, Lattice-tiling by certain star bodies, *Studia Sci. Math. Hung.* **20** (1985), 71–76.

14. S. Szabó, On mosaics consisting of multidimensional crosses, *Acta Math. Acad. Sci. Hung.* **38** (1981), 191–203.

15. ——, Finite abelian groups and n-dimensional mosaics (Hungarian), *Mat. Lapok.* **28** (1977/1980), 305–318 (MR 82i:20065).

16. ——, A bound of k for tiling by (k, n)-crosses and semicrosses, *Acta Math. Acad. Sci. Hung.* **44** (1984), 97–99.

17. ——, A star polyhedron that tiles but not as a fundamental domain, *Colloquia Math. Soc. János Bolyai* **48**, Siófok, (1985).

18. W. Ulrich, Non-binary error correction codes, *The Bell System Technical Journal* (1957), 1341–1388.

19. A. J. Woldar, A reduction theorem on purely singular splittings of cyclic groups, *Proc. Amer. Math. Soc.,* to appear.

Chapter 4
Packing and Covering by the Semicross and Cross

When a cluster C does not tile n-space, we immediately face two questions: How densely can we pack translates of C? How thinly can we place translates of C such that they cover all of n-space? Similar questions for the n-dimensional ball, going back to Kepler, have been answered only for small values of n. We will get good estimates in all dimensions in the case of Z-lattice packings by semicrosses and even much more information about packing by crosses. Similar questions about Z-lattice coverings seem to be far more difficult, as we will see later in the chapter.

In order to quantify the efficiency of a packing or covering we need the notion of the "density" of a family of translates of a cluster.

Let C be a cluster in n-space consisting of q cubes and $S = \{a_1, a_2, \ldots\}$ a denumerable sequence of points in n-space without an accumulation point. For each positive number r let $Q(r)$ be the cube $\{(x_1, x_2, \ldots, x_n) : |x_i| \le r\}$, which has side $2r$ and volume $(2r)^n$. Let $N(r)$ be the number of points a_i in S such that $a_i + C$ lies in $Q(r)$. The fraction of $Q(r)$ filled by these translates is

$$f(r) = \frac{qN(r)}{(2r)^n}.$$

(In case of a packing $f(r) \le 1$.) Assume that $f(r)$ approaches a limit as $r \to \infty$ and denote this limit $d(S)$, which we call the *density* of the family of translates.

Exercise 1. Prove that the density of a covering is at least 1.

We may think of the density of a family of translates of a cluster as the average number of times that points in n-space are covered by the translates.

The *packing density* of the cluster C, usually denoted $\delta(C)$, is defined as the least upper bound of the densities of packings by C. The *covering density* of C, denoted $\theta(C)$, is defined as the greatest lower bound of the densities of coverings by C. If in these definitions we restrict the families to be lattices of translates of C, then we obtain the *lattice packing density,* $\delta_L(C)$, and *lattice covering density* $\theta_L(C)$. Note that

$$\delta_L(C) \le \delta(C) \le 1 \le \theta(C) \le \theta_L(C).$$

If we restrict the translating vectors to be an integer lattice, then we obtain the *Z-lattice packing density* $\delta_L^Z(C)$ and *Z-lattice covering density* $\theta_L^Z(C)$.

Analogous densities have also been defined when C is a convex set and have been studied for over two centuries when C is a ball. References for what has been discovered are [1, pp. 15–17], [6, p.3], or [9, pp. 177–185]. For $n = 1, 2$, and 3 the densest packing of balls is achieved with a lattice packing; a proof for $n = 3$, due to Hsiang, was obtained as recently as 1991, and its details are still being checked.

Any denumerable packing of n-space by translates of a cluster C can be shifted locally to produce another such packing in which all the coordinates of the translating vectors are integers [3]. This shift, which generalizes the idea of "rounding down," is defined as follows.

Consider a denumerable family of translates of C that packs n-space. Each cube in these clusters has the form $w + A$, where

$$A = \{(x_1, \ldots, x_n) : 0 \le x_i \le 1,\ 1 \le i \le n\}.$$

Let W be the set of w's.

Since W is denumerable, we may translate the family W by some vector c so that no vector $c+w$ with $w \in W$ is on the boundary of $z+A$ for any $z \in Z^n$. Assume that this initial shift has been made.

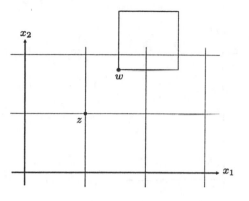

FIGURE 1

Then each w is in the interior of a unique cube of the form $z + A$ where $z \in Z^n$. Call $z + A$ the *shift* of $w + A$ and z the *shift* of w. See Figure 1.

Note that the set of shifts of the cubes in a cluster form a translation of that cluster. So the shift can be viewed as translating clusters. Moreover, the density of the family obtained by the shift is the same as the density of the original family.

Assuming that the family $\{w + A : w \in W\}$ is a packing, we will show that the shift of this family is also a packing.

Let $w_1 + A$ and $w_2 + A$ be distinct cubes in the packing, and $z_1 + A$ and $z_2 + A$ their respective shifts. If the interiors of $z_1 + A$ and $z_2 + A$ have a nonempty intersection, then $z_1 = z_2$. Thus $w_1 = a_1 + z_1$ and $w_2 = a_2 + z_2$, where a_1 and a_2 are in the interior of A. Hence $w_1 + a_2 = w_2 + a_1$, which violates the assumption that cubes $w_1 + A$ and $w_2 + A$ are part of a packing.

To show that the shift of a covering is again a covering, consider a cube $z + A$, $z \in Z^n$. We will show that there is a cube $w + A$ in the covering whose shift is $z + A$. Therefore, the densest packing by a cluster has the same density as the densest integer packing.

The point $z + (1, 1, \ldots, 1)$ lies in the interior of $w + A$ for some $w \in W$, as illustrated in Figure 2. Then it is not hard to show that $z + A$ is the shift of $w + A$.

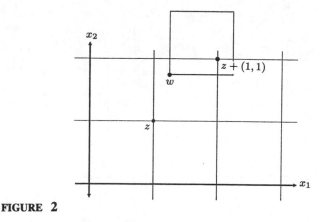

FIGURE 2

Exercise 2.   Justify the last sentence.

The shift of a lattice is not necessarily a lattice. (Recall that the cluster formed of two squares separated by one square lattice tiles 2-space but does not Z-lattice tile 2-space.)

The next exercise provides two more illustrations of this fact.

Exercise 3.
(a) Let $a > 1$ be an irrational number. Translates of the interval $[0, 1]$ by the numbers $\{na : n \in Z\}$ form a lattice packing of R^1. Show that the shift of this packing is not a lattice packing.
(b) Consider any lattice packing of R^n by cluster C where the determinant of the lattice is irrational. Show that its shift is not a lattice.

1. Packing by semicrosses

Let C be the (k, n)-semicross, which has volume $kn + 1$. A packing of R^n with translates of the (k, n)-semicross by vectors in the lattice H of determinant d therefore has density

$$\frac{kn + 1}{d}.$$

Now restrict H to being a sublattice of Z^n. The family of translates of the (k, n)-semicross by translates of the vectors of H is a packing if and only if the cosets $\{ie_j + H : 1 \le i \le k, 1 \le j \le n\}$ are distinct and distinct from the coset H. Let $G = Z^n/H$. Then there are n elements in G, $s_1, s_2, \ldots, s_n$, such that the kn elements is_j, $1 \le i \le k, 1 \le j \le n$ are distinct and nonzero.

Exercise 4. Justify the last two statements.

Finding the densest Z-lattice packing by the (k, n)-semicross is equivalent to finding the smallest group G such that there are n elements in G, $s_1, s_2, \ldots, s_n$, where the kn elements is_j, $1 \le i \le k, 1 \le j \le n$, are distinct and distinct from 0. If such elements $s_1, s_2, \ldots, s_n$ exist in G we will say that $S(k)$ then n-packs G and that $\{s_1, s_2, \ldots, s_n\}$ is a *packing set*.

Exercise 5. Show that for every pair of positive integers k and n there is a finite abelian group that $S(k)$ n-packs.

Let $g(k, n)$ be the order of the smallest abelian group G that $S(k)$ n-packs. Obviously, $g(k, n) \ge kn + 1$ and equals $kn + 1$ only if $S(k)$ splits an abelian group of order $kn + 1$. Moreover, for fixed n, $g(k, n)$ is a nondecreasing function of k. The density of the densest Z-lattice packing of R^n by the (k, n)-semicross is

$$\frac{kn + 1}{g(k, n)}.$$

In dimensions 1 and 2, $g(k, n)$ is easy to determine: $g(k, 1) = k + 1$ and $g(k, 2) = 2k + 1$. We will show that

$$\lim_{k \to \infty} \frac{g(k, 3)}{k^{3/2}} = 1. \tag{1}$$

The proof is based on two lemmas [7]. The first shows that $g(k, 3) \ge (k + 1)^{3/2}$.

Lemma 1. *Let G be an abelian group of order m such that $S(k)$ 3-packs G. Then $(k + 1)^3 \le m^2$.*

Proof. Let the packing set be $\{a, b, c\}$. Let $G \times G$ be the set of ordered pairs (u, v), where u and v are in G. Consider the $(k + 1)^3$ elements $(xa - yb, yb - zc)$, $0 \leq x, y, z \leq k$, in the set $G \times G$. If $(k + 1)^3 > m^2$, then by the pigeon-hole principle, two of these elements coincide, say

$$xa - yb = \overline{x}a - \overline{y}b \quad \text{and} \quad yb - zc = \overline{y}b - \overline{z}c,$$

$$0 \leq x, \overline{x}, y, \overline{y}, z, \overline{z} \leq k, \qquad (x, y, z) \neq (\overline{x}, \overline{y}, \overline{z}).$$

If $x = \overline{x}$, then $yb = \overline{y}b$. Since $0 \leq y, \overline{y} \leq k$ and b is an element in the packing set $\{a, b, c\}$, $y = \overline{y}$. Similarly, $z = \overline{z}$. But this contradicts the fact that (x, y, z) and $(\overline{x}, \overline{y}, \overline{z})$ are distinct. Similarly, $y \neq \overline{y}$ and $z \neq \overline{z}$.

Without loss of generality, we may assume that $x > \overline{x}$. Then we have $(x - \overline{x})a = (y - \overline{y})b$. Since a and b belong to a packing set, $y - \overline{y}$ is negative. The equation $(\overline{y} - y)b = (\overline{z} - z)c$ then implies that $\overline{z} - z$ is negative, hence $z > \overline{z}$. We have

$$(x - \overline{x})a = (y - \overline{y})b = (z - \overline{z})c,$$

hence $(x - \overline{x})a = (z - \overline{z})c$, which contradicts the assumption that $\{a, b, c\}$ is a packing set. Thus $(k + 1)^3 \leq m^2$, and the lemma is proved. $\square$

The next lemma puts a lower bound on $g(k, 3)$ for an infinite set of integers k.

Lemma 2. *Let $a \geq 2$ be an integer and let $k = a^2 - a$. Then $S(k)$ 3-packs the cyclic group $C(a^3 + 1)$ with packing set $\{1, -a, a^2\}$.*

Proof. The proof that the congruence $i \cdot 1 \equiv j(-a) \pmod{a^3 + 1}$, has no solutions for $1 \leq i, j \leq a^2 - a$ is straightforward. The congruence $i \cdot 1 \equiv j(a^2) \pmod{a^3 + 1}$ reduces to the previous one by multiplying by $-a$. The congruence $i(-a) \equiv ja^2 \pmod{a^3 + 1}$ also reduces to the first one by division by a (which is relatively prime to the modulus, $a^3 + 1$). $\square$

Incidentally, for $a = 2$ and 3 and $k = a^2 - a$, $g(k, 3)$ equals $a^3 + 1$. It is not known whether this true for all a.

Exercise 6. Fill in the details of the preceding proof.

Lemmas 1 and 2 imply the following theorem.

Theorem 1.

$$\lim_{k \to \infty} \frac{g(k,3)}{k^{3/2}} = 1.$$

Outline of proof. By Lemma 2, for $k = a^2 - a$, $g(k,3) \le a^3 + 1 = (1 + \sqrt{1 + 4k})^3/8 + 1$. Thus, for k of the form $a^2 - a$,

$$(k+1)^{3/2} \le g(k,3) \le \frac{(1 + \sqrt{1 + 4k})^3}{8} + 1.$$

Therefore, for these values of k,

$$\lim_{k \to \infty} \frac{g(k,3)}{k^{3/2}} = 1. \tag{2}$$

To treat an arbitrary k, let a be the positive integer such that

$$a^2 - a \le k < (a+1)^2 - (a+1).$$

Let $k_1 = a^2 - a$ and $k_2 = (a+1)^2 - (a+1)$. Then

$$g(k_1, 3) \le g(k, 3) \le g(k_2, 3).$$

Note that

$$\lim_{k \to \infty} \frac{k_1}{k_2} = 1. \tag{3}$$

Equations (2) and (3) imply that (1) holds when k is not restricted to be of the form $a^2 - a$. $\qquad\square$

Exercise 7. Fill in the details in the preceding proof.

Problem 1. If $S(k)$ n-packs the finite abelian group G, must it n-pack the cyclic group $C(|G|)$? (Theorem 9 in the preceding chapter shows that if the packing is a splitting of G, then the answer is "yes.")

Exercise 8.
(a) Show that for a positive odd integer b, $S((b^2 - 1)/2)$ 4-packs the cyclic group $C((b + 1)(b^2 + 1)/2)$ with the splitting set

$\{1, -b, (-b)^2, (-b)^3\}$. (Hint: first show that the packing set is a subgroup of $C^*((b+1)(b^2+1)/2)$.

(b) Deduce that

$$\limsup_{k \to \infty} \frac{g^2(k, 4)}{k^3} \le 2.$$

Exercise 9.

(a) Show that for $b \equiv 1 \pmod 6$ and at least 7, $S((b^2 + b - 2)/3)$ 6-packs $C((b^2 + b + 1)(b+1)/3)$ with packing set

$$\{1, -b, (-b)^2, (-b)^3, (-b)^4, (-b)^5\}.$$

(b) Deduce that

$$\limsup_{k \to \infty} \frac{g^2(k, 6)}{k^3} \le 3.$$

In Theorem 1 and Exercises 8 and 9 the method rests on the fact that $x^3 - 1$, $x^4 - 1$, and $x^6 - 1$ have cubic factors with certain properties. The number k is expressed as a quadratic in b, and for this k, $S(k)$ is shown to pack a cyclic group whose order is a cubic in b. (The packing set is generated by $-b$ and forms a group under multiplication.) In these cases

$$\limsup_{k \to \infty} \frac{g^2(k, n)}{k^3}$$

is shown to be no greater than some specific rational number. However, it is proved in [5] that for $n \ge 3$

$$\lim_{k \to \infty} \frac{g^2(k, n)}{k^3} = 4 \cos^2\left(\frac{\pi}{n}\right). \tag{4}$$

As Exercise 10 shows, for $n \ge 3$ this limit is rational only when $n = 3, 4$, or 6. This means that the method described has to be modified to treat general n. The key modification to obtain a proof of (4) is to consider packing sets of the form $\{1, -b, (-b)^2, \ldots, (-b)^{n-1}\}$ that do not necessarily form a group. In other words, do not demand that $(-b)^n = 1$.

Exercise 10. Let n be an integer and ω a primitive nth root of unity. Assume that $[Q(\omega) : Q] = \phi(n)$, where ϕ is the Euler "ϕ function."

(a) Show that $[Q(\omega) : Q(\cos{(2\pi/n)}] = 2$ for $n \geq 3$.
(b) Show that $Q(\cos^2{(\pi/n)}) = Q(\cos{(2\pi/n)})$ for $n \geq 1$.
(c) Show that for $n \geq 3$, $\cos^2{(\pi/n)}$ is rational if and only if $n = 3, 4,$ or 6.

Let us return to the case $n = 3$. Since

$$\lim_{k \to \infty} \frac{g(k, 3)}{k^{3/2}} = 1,$$

we can calculate how well the $(k, 3)$-semicross Z-lattice packs R^3. Letting $\delta_L^Z(k)$ be the density of the densest such packing, we have $\delta_L^Z(k) = (3k + 1)/g(k, 3)$. So we have

$$\lim_{k \to \infty} \frac{\delta_L^Z(k)}{3/\sqrt{k}} = 1. \tag{5}$$

Speaking loosely, we can say that when k is large, the densest Z-lattice packing of the $(k, 3)$-semicross fills only about $\frac{3}{\sqrt{k}}$ of R^3. This has been generalized to arbitrary lattice packings in [7].

If we remove the condition that the packing be a lattice packing, it turns out that the $(k, 3)$-semicross packs R^3 much more densely. In fact, if $\delta(k)$ denotes the packing density of the $(k, 3)$-semicross, we have

$$\lim_{k \to \infty} \frac{\delta_L^Z(k)}{\delta(k)} = 0. \tag{6}$$

The combinatorial argument [4] rests on the construction of certain designs, called "monotonic matrices," which we now describe.

Consider a cube of width k, which is composed of k^3 unit cubes. Say that you place $t(k)$ disjoint translates of the $(k, 3)$-semicross in such a way that each "corner cube" of a semicross coincides with one of the k^3 unit cubes. Then these semicrosses lie in the cluster

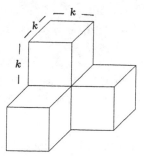

FIGURE 3

formed of four copies of the cube of width k, which is a magnification of a $(1, 3)$-semicross, as shown in Figure 3.

Of the $4k^3$ unit cubes in this cluster, $t(k)(3k + 1)$ are occupied
by the $t(k)$ semicrosses. Since the $(1, 3)$-semicross tiles Z^3, there
would be a packing of R^3 with density

$$\frac{t(k)(3k + 1)}{4k^3}.$$

Identify each of the k^3 unit cubes in the cube of width k with
a triple (x, y, z), the coordinates of its center. Then $x, y,$ and z are
integers, and $1 \le x, y, z \le k$. Because the semicrosses are disjoint,
for a given (x, y) there is at most one triple (x, y, z). We can therefore record the configuration with the aid of a square array of k^2
cells in which the cell corresponding to (x, y) holds the number z if
the semicross with corner (x, y, z) is present. Otherwise cell (x, y)
is empty. This leads us to consider certain square arrays, which we
call "monotonic matrices."

A *monotonic matrix* of order k is an array of k^2 cells in some of
which an integer in the set $\{1, 2, \ldots, k\}$ is placed, subject to three
conditions:

(a) For two filled-in cells in a row, the one further right has a larger
    entry;

(b) For two filled-in cells in a column, the higher one has the larger
    entry;

(c) For two filled-in cells with the same entry, the one further right is higher (the "positive slope" condition).

Exercise 11.   Show that conditions (a), (b), and (c) are equivalent to the fact that the corresponding semicrosses are disjoint.

Let $m(k)$ be the maximum number of occupied cells in all monotonic matrices of order k. Note that the $(k, 3)$-semicross tiles Z^3 with density at least $m(k)(3k + 1)/4k^3$.

Table 1 lists some of the values of $m(k)$; numbers in parentheses are lower bounds. For instance, that $m(3) \geq 5$ is shown by the array in Figure 4.

K. Joy programmed an exhaustive search for the largest monotonic matrix of order 5, determining that $m(5) = 11$. Figure 5 displays one of the many arrays he found.

Exercise 12.   Show that $m(2) = 2$, $m(3) = 5$, and $m(4) \geq 8$.

Exercise 13.   Show that if k is a square, then $m(k) \geq k^{3/2}$. (Table 1 shows that $m(4) = 4^{3/2}$ but $m(9) > 9^{3/2}$.)

Exercise 14.   Show that $m(k \cdot l) \geq m(k)m(l)$
 (a) by using the geometry of semicrosses;
 (b) by using only the definition of a monotonic matrix.

k	1	2	3	4	5	6	7	8	9	10
$m(k)$	1	2	5	8	11	(14)	(19)	(22)	(28)	(32)

TABLE 1

2		3
		1
1	3	

FIGURE 4

		4		5
4			5	
		1	2	3
3	5			
1	2			

FIGURE 5

By Table 1, $m(7) \geq 19 \approx 7^{1.513}$. Therefore, by Exercise 14, $m(7^r) \geq (m(7))^r \geq (7^r)^{1.513}$, for any positive integer r. Thus for $k = 7^r$ we have $m(k) \geq k^{1.513}$ and there are packings by the $(k, 3)$-semicross with density at least

$$\frac{k^{1.513}(3k + 1)}{4k^3}. \tag{7}$$

Comparing (7) with (5) establishes (6). The key is that 1.513 is greater than 1.5.

Exercise 15. Show that $m(k + 1) > m(k)$.

In a moment we will apply the next exercise to the function $m(k)$.

Exercise 16. Let $f(n)$ be a positive real number for each positive integer n. Assume that $f(n + 1) > f(n)$, $f(n) \leq n^2$, and $f(mn) \geq f(m)f(n)$, for any positive integers m, n. Define e_n by the equation $f(n) = n^{e_n}$. Show that

$$\lim_{n \to \infty} e_n$$

exists, and is at most 2. (Hint: It is the least upper bound of $\{e_n\}$.)

By Exercise 16, if we define e_k by setting $m(k) = k^{e_k}$, we know that

$$\lim_{k \to \infty} e_k$$

exists. Call it L. Since $e_7 \geq 1.513$, we have $1.513 \leq L \leq 2$.

Problem 2.   Find L.

Problem 3.   Examine $m(k)/k^2$ as $k \to \infty$. If $L < 2$, then $m(k)/k^2 \to 0$. (Does the limit exist?)

Hickerson pointed out in conversation that if we knew that the density of the densest packing of R^3 by the $(k, 3)$-semicross approaches 0 as $k \to \infty$, we would immediately have the analogous result for all dimensions $n > 3$. To see this, assume that for some dimension $n > 3$ there are arbitrarily large integers k such that the (k, n)-semicross packs R^n with density greater than ε, a fixed positive number. Using the shift, we may assume that the translating vectors lie in Z^n.

The number of corner cubes of the semicrosses in a packing in some large cube of side s (which we take to be an integer much larger than k) would then be greater than $\varepsilon s^n/(nk + 1)$. Since the large cube could be viewed as the union of s "$(n - 1)$-dimensional" slabs of unit thickness, in some $(n - 1)$-dimensional cross sectional slab of this large cube there would be at least $\varepsilon s^{n-1}/(nk+1)$ corner cubes.

Now, the cluster similar to the $(1, n - 1)$-semicross, but composed of n cubes of side s tiles R^{n-1}. Of its ns^{n-1} unit cubes at least $\varepsilon s^{n-1}((n - 1)k + 1)/(nk + 1)$ are occupied by the $(1, n - 1)$-semicrosses whose corners lie in the corner of the large semicross. Therefore there is a packing of R^{n-1} by the $(1, n - 1)$-semicross of density at least

$$\frac{\varepsilon s^{n-1}((n - 1)k + 1)}{ns^{n-1}(nk + 1),}$$

which approaches $\varepsilon(n - 1)/n^2$ as $k \to \infty$. Repeating this descending argument until we reach $n = 3$ produces a contradiction of our assumption that the packing density of the $(k, 3)$-semicross approaches 0 as $k \to \infty$.

2. Packing by crosses

The case of Z-lattice packings of R^n by the (k, n)-cross is quite different. For $n \geq 2$, let $h(k, n)$ be the order of a smallest abelian group that $F(k)$ n-packs. It is known [7] that

$$\lim_{k \to \infty} \frac{h(k, n)}{k^2} = 1. \tag{8}$$

One difference between (4) and (8) is that the dimension n does not influence (8) but does influence (4). The other is that the exponent of k is 2 instead of $3/2$, which implies that the densest Z-lattice packing of R^n by (k, n)-crosses is much less dense than is the case for semicrosses. If $\delta^*(k)$ is that density,

$$\lim_{k \to \infty} \frac{\delta^*(k)}{(2kn + 1)/k^2} = 1,$$

hence

$$\lim_{k \to \infty} \frac{\delta^*(k)}{2n/k} = 1. \tag{9}$$

Another difference between the cross and semicross is that the proof of (8) uses packing sets that form arithmetic progressions. We will not present this proof, but will content ourselves by showing that for $n \geq 2$, $h(k, n) \geq (k + 1)^2$.

Assume that $F(k)$ n-packs the finite abelian group G. Then $F(k)$ 2-packs G with packing set $\{a, b\}$. Assume that $|G| < (k+1)^2$. Consider the $(k+1)^2$ elements $ia + jb, 0 \leq i, j \leq k$. By the pigeon-hole principle, there are distinct pairs (i, j) and $(\bar{i}, \bar{j}), 0 \leq i, \bar{i}, j, \bar{j} \leq k$ such that $ia + jb = \bar{i}a + \bar{j}b$, or equivalently $(i - \bar{i})a = (\bar{j} - j)b$. Note that $|i - \bar{i}| \leq k$ and $|\bar{j} - j| \leq k$. If $i = \bar{i}$, we have $(\bar{j} - j)b = 0$, contradicting the fact that b is part of a set that packs G. Thus $i - \bar{i} \neq 0$ and, similarly, $\bar{j} - j \neq 0$. This violates the assumption that $\{a, b\}$ is a packing set. Consequently $h(k, n) \geq (k + 1)^2$ for all $n \geq 2$.

Exercise 18.
(a) Show that for even k, $F(k)$ 2-packs $C(k^2 + 2k + 2)$ with packing set $\{1, k + 1\}$.
(b) Show geometrically that $h(k, 2) = k^2 + 2k + 2$.

Exercise 19.   Show that $F(k)$ 3-packs $C(k^2 + 3k + 3)$ with packing set $\{1, k + 1, k + 2\}$.

Exercise 20. Show that for even k, $F(k)$ 4-packs $C(k^2 + 4k + 5)$ with packing set $\{1, k + 1, k + 2, k + 3\}$.

Exercise 21. Show that for $k \equiv 0$ or $4 \pmod 6$, $F(k)$ 5-packs $C(k^2 + 8k + 9)$ with packing set $\{1, k + 1, k + 3, k + 5, k + 7\}$.

Exercise 22. Show that for $k \equiv 0 \pmod 6$, $F(k)$ 6-packs $C(k^2 + 30k + 233)$ with packing set $\{1, k+11, k+13, k+15, k+17, k+19\}$.

Problem 4. Determine $h(k, n)$. Trivially, $h(k, 1) = 2k + 1$.

Unlike the semicross case, the ratio between the density of the densest Z-lattice packing and densest packing does not approach 0 as k approaches infinity. Instead, it approaches 1. The argument is not long.

First of all, for any packing of R^n by the (k, n)-cross, the shift produces a packing of the same density. So we consider only integer packings. In a square tray of side length $k + 1$ there can be at most one center of a cross in a packing. Since these trays tile R^n, the density of the packing is at most $(2kn + 1)/(k + 1)^2$, which is asymptotic to $2n/k$ as $k \to \infty$. Comparing this with (8) shows that Z-lattice packings by the (k, n)-cross, when k is large, can be just about as dense as an arbitrary packing by the (k, n)-cross.

Problem 5. Find the density of the densest packing of R^n by the (k, n)-cross.

Problem 6. Is the density of the densest packing of R^n by a cluster always a rational number?

3. Covering by the semicross and cross

If a semicross or a cross does not tile R^n, we may also ask, "How thinly can it cover R^n?"

We consider only covers of Z^n, and view a semicross or cross as subset of Z^n, and deal only with translates by integer vectors.

Such a family of translates of a semicross or cross is said to *cover Z^n* if every point in Z^n is in at least one of the translates. The

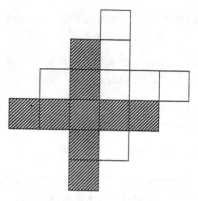

FIGURE 6

density of the covering can be viewed as "the average number of times each point is Z^n is covered."

For instance, the $(2,2)$-cross covers R^2 with density $9/8$, since two overlapping copies of this cross fill up the cluster of 16 squares shown in Figure 6, and this latter cluster tiles the plane [3]. Hence the $(2,2)$-cross covers the plane with density $18/16 = 9/8$.

Exercise 23. Verify that the cluster in Figure 6 tiles the plane.

Exercise 24. Use a similar approach to get a covering of the plane by the $(3,2)$-cross of density $13/11$.

The construction of Z-lattice coverings of R^n by the (k,n)-semicross, just like Z-lattice tilings or packings, depends on finite groups. We say that $S(k)$ *n-covers* a finite abelian group G with the *covering set* $\{s_1, s_2, \ldots, s_n\}$ if every element in $G \setminus \{0\}$ can be expressed in the form is_j, $1 \leq i \leq k$, $1 \leq j \leq n$. Let $f(k,n)$ be the order of the largest abelian group that $S(k)$ n-covers. Clearly $f(k,n) \leq kn+1$. Just as we can obtain a lattice packing of Z^n from a packing of a finite group, we can obtain a lattice covering of Z^n from a covering of a finite group.

Exercise 25. Show that the density of the least dense Z-lattice covering of Z^n by the (k,n)-semicross is $(kn+1)/f(k,n)$.

Determining the behavior of $f(k, n)$ for large k is much harder than determining the behavior of $g(k, n)$, the corresponding function for packings. In the case of a packing, if g_1 and g_2 are two elements of a packing set we can say that the equation $ig_1 = jg_2$, $1 \leq i, j \leq k$ has no solution. However, for coverings, the structure is looser; there is nothing of use that we can say about a pair of elements in a covering set. Even determining $f(k, 3)$ is far from trivial.

The next theorem provides very efficient coverings by semicrosses in $(p + 1)$-space, where p is a prime. It, as well as the other results on covering that we will describe, are due to Dad-del [2].

Theorem 2. *Let p be a prime and r be a positive integer. Let $k = rp - 1$, $m = rp^2$, and $n = p + 1$. Then $S(k)$ n-covers $C(m)$ with the covering set $\{p\} \cup \{1 + irp : 0 \leq i \leq p - 1\}$.*

Proof. Consider an integer $x \in [1, m - 1]$. If p divides x then x is of the form ip, where $i \in S(k)$.

Let $A = \{x : 1 \leq x \leq m, (x, p) = 1\}$. We will show that A is covered. For convenience, if $g \in C(m)$, let $\overline{g}$ denote $S(k)g$, "the set of elements covered by g."

For each i, $0 \leq i \leq p - 1$,

$$|A \cap \overline{1 + irp}| = k - [k/p] = r(p - 1).$$

Since $|A| = rp(p - 1)$, all that remains is to show that if $i \neq i'$, then $\overline{1 + irp} \cap \overline{1 + i'rp} = \varnothing$.

Assume that there are integers j, j', $1 \leq j, j' \leq k$ such that

$$j(1 + irp) \equiv j'(1 + i'rp) \pmod{rp^2}.$$

Then $j \equiv j' \pmod{rp}$, hence $j = j'$. Consequently

$$jirp \equiv ji'rp \pmod{rp^2}$$

and we have

$$j(i - i') \equiv 0 \pmod{p}.$$

Since we assume that $j(1+irp)$ is not divisible by p, $(j,p) = 1$. Thus $i - i' \equiv 0 \pmod{p}$, from which it follows that $i = i'$. This completes the proof.   □

It can be shown that the covering set in the preceding theorem is essentially unique.

In any case, the covering is quite economical. The $m - 1 = rp^2 - 1$ nonzero elements of $C(m)$ are covered by kn products, and $kn = (rp - 1)(p + 1) = rp^2 + (r - 1)p - 1$, which is not much larger (in ratio) than $rp^2 - 1$. Holding n fixed, we have, since $k = rp - 1$

$$\limsup_{k \to \infty} \frac{f(k, n)}{k} \geq \limsup_{r \to \infty} \frac{rp^2}{rp - 1} = p.$$

The proof of Theorem 2 uses only cyclic groups. This suggests the following problem.

Problem 7. If $S(k)$ n-covers a finite abelian group G, does it n-cover $C(|G|)$? (For $k = 2$ the answer is "yes" [8].)

Theorem 2 implies that when the dimension n is one more than a prime and k is large there are Z-lattice coverings of R^n by the (k, n)-semicross with density near $n/(n - 1)$, which for large n is near 1. This is quite a contrast with the fact that Z-lattice packings by these semicrosses have densities near 0.

The construction of Z-lattice coverings by the (k, n)-cross reduces to the following problem on finite abelian groups. We say $F(k) = \{\pm 1, \pm 2, \ldots, \pm k\}$ n-covers a finite abelian group G with covering set $\{s_1, s_2, \ldots, s_n\}$ if every element of $G \setminus \{0\}$ can be expressed in the form is_j, $1 \leq |i| \leq k$, $1 \leq j \leq n$. Let $c(k, n)$ be the order of the largest abelian group that $F(k)$ n-covers. Clearly $c(k, n) \leq 2kn + 1$. An argument similar to that for Theorem 2 yields the following theorem.

Theorem 3. *Let r be an even positive integer, q prime, $k = rq/2$, $n = q + 1$, and $m = rq^2$. Then $F(k)$ n-covers $C(m)$ with covering set* $\{q\} \cup \{1 + irq : 0 \leq i \leq q - 1\}$.

Exercise 26. Prove Theorem 3.

Returning to the covering problem suggested by semicrosses, we can easily check that $f(k, 1) = k + 1$ and $f(k, 2) = 2k + 1$. We might expect $f(k, 3)$ to be approximately $3k$. If so, we are in for a surprise. Very little is gained by adjoining another element to the 2-covering set. In fact $f(k, 3) = 2k + 2$.

Exercise 27. Show that $S(k)$ 3-covers $C(2k + 2)$ with covering set $\{1, -1, k + 1\}$.

We thus have

$$\lim_{k \to \infty} \frac{f(k, 1)}{k} = 1, \qquad \lim_{k \to \infty} \frac{f(k, 2)}{k} = 2, \qquad \lim_{k \to \infty} \frac{f(k, 3)}{k} = 2.$$

For $n \geq 4$, $\lim_{k \to \infty} f(k, n)/k$ has not been determined. It has not even been shown to exist. As noted, when $n = p + 1$ for a prime p,

$$\limsup_{k \to \infty} \frac{f(k, n)}{k} \geq p.$$

Theorem 3, though it concerns the cross, also provides information about the semicross, since when $F(k)$ n-covers a group G with the covering set $\{s_1, s_2, \ldots, s_n\}$, $S(k)$ $2n$-covers G with the covering set $\{\pm s_1, \pm s_2, \ldots, \pm s_n\}$.

For instance, consider Theorem 3 in the case $q = 7$ (and $2n$ is therefore 16). According to the theorem, when r is even, $S(7r/2)$ 16-covers $C(49r)$. Thus

$$\limsup_{k \to \infty} \frac{f(k, 16)}{k} \geq 14.$$

Theorem 2 gives a weaker result: The case $p = 13$ gives us

$$\limsup_{k \to \infty} \frac{f(k, 16)}{k} \geq 13.$$

Incidentally, $n = 16$ is the smallest dimension for which Theorem 3 gives a stronger result for the function f than does Theorem 2.

Exercise 28. For which values of $n \leq 50$ does Theorem 3 give a stronger result than does Theorem 2?

On the basis of extensive computations in low dimensions as well as Theorems 2 and 3, it is tempting to conjecture that, for $n \geq 3$,

$$\lim_{k \to \infty} \frac{f(k, n)}{k}$$

is either the largest prime less than n or twice the largest prime less than the greatest integer less than $n/2$, whichever is larger. However, it should be kept in mind that we do not even know that the limit exists.

Problem 8. Does $\lim_{k \to \infty} \frac{f(k,n)}{k}$ exist? If so, what is its value?

Problem 9. Settle for the cross the analog of Problem 8.

Exercise 29. Show that $S(k)$ 4-covers $C(m)$ with covering set $\{\pm 1, \pm 2\}$:
(a) if k is even and $m = 3k + 1$;
(b) if k is odd and $m = 3k + 2$.

Exercise 30. Show that for $k \neq 2$, $S(k)$ 5-covers $C(m)$ with the covering set given as follows (in the case $k = 6l + 1$, x is arbitrary):

k	m	covering set
$6l + 0$	$3k + 3$	$\{\pm 1, \pm 2, k + 1\}$
$6l + 1$	$3k + 2$	$\{\pm 1, \pm 2, x\}$
$6l + 2$	$3k + 5$	$\{1, -2, \pm 3, -6\}$
$6l + 3$	$3k + 4$	$\{1, -2, -3, \pm 6\}$
$6l + 4$	$3k + 5$	$\{1, -2, -3, \pm 6\}$
$6l + 5$	$3k + 4$	$\{1, -2, -3, \pm 6\}$

If the conjecture made after Exercise 28 is correct, we would have the following table:

n	1	2	3	4	5	6	7	8	9	10	11	12	13	14	15	16	17	18
$\lim_{k \to \infty} \dfrac{f(k, n)}{k}$	1	2	2	3	3	5	5	7	7	7	7	11	11	13	13	14	14	17

It would imply that while there are very efficient Z-lattice coverings of R^{12} by the $(k, 12)$-semicross for large k, the situation in R^{11} is quite different, where the most efficient Z-lattice coverings have density approximately $11/7 \approx 1.57$. For large n this density approaches 1, which implies that R^{11} is the space that is "hardest" to Z-lattice cover by semicrosses.

Up to this point in the book we have considered only tilings, packings, and coverings by clusters. Even these simple objects, composed of a finite number of cubes, even of a single cube, raise a host of geometric, algebraic, and combinatorial problems, most of which are unsolved. In the next two chapters we investigate tilings by triangles. Chapter 5 is concerned with tiling a polygon with triangles all of which have the same area. On the other hand, Chapter 6 examines tilings by triangles where there is some restriction on their shapes, but none on their areas. For instance, it will show that a square cannot be tiled by $30° - 60° - 90°$ triangles. As is to be expected, the algebraic techniques used in the coming two chapters are quite different. To treat triangles of equal areas we use valuation theory, but to deal with triangles of prescribed shapes we will use isomorphisms of subfields of the complex numbers.

References

1. J. H. Conway and N. J. A. Sloane, *Sphere packings, lattices, and groups*, Springer-Verlag, New York, 1988.

2. A. Dad-del, Covering abelian groups with cyclic subsets, *Aequationes Math.*, 1994 (to appear).

3. H. Everett and D. Hickerson, Packing and covering by translates of certain nonconvex bodies, *Proc. Amer. Math. Soc.* **75** (1979), 87–91.

4. W. Hamaker and S. K. Stein, Combinatorial packing of R^3 by certain error spheres, *IEEE Trans. Inf. Theory.* IT–30 (1984), 364–368.

5. D. R. Hickerson and S. K. Stein, Abelian groups and packing by semicrosses, *Pacific J. Math.* **122** (1986), 95–109.

6. C. A. Rogers, *Packing and Covering*, Cambridge University Press, Cambridge, 1964.

7. S. K. Stein, Packing of R^n by certain error spheres, *IEEE Trans. Inf. Theory,* IT–30 (1984), 356–363.

8. S. Szabó, Lattice covering by semi-crosses of arm length two, *European J. of Comb.* **12** (1991), 263–266.

9. T. Thompson, *From error correcting codes through sphere packings to simple groups*, Mathematical Association of America, Washington, D. C., 1983.

Chapter 5
Tiling by Triangles of Equal Areas

In the first four chapters we examined tilings of Euclidean space by translates of sets that are unions of cubes. In this chapter we consider tilings of a much different character, namely tilings of polygons by triangles of equal areas. (We do not demand that the triangles be congruent.) We restrict our attention to simplicial dissections, though the theorems hold for all dissections.

In 1965 Fred Richman at the University of New Mexico at Las Cruces wanted to include a geometric question in a master's degree examination he was preparing. "I noticed that it was easy to cut a square into an even number of triangles of equal areas, but I could not see how to cut it into an odd number." Not solving his problem, he chose not to put it on the examination. However, he did show that the square could not be cut into three or five triangles of equal areas and mentioned the problem to his colleague and bridge partner, John Thomas. Thomas recalls that

> Everyone to whom the problem was put (myself included) said something like 'that is not my area but the question surely must have been considered and the answer is probably well known.' Some thought they had seen it, but could not remember where. I was interested because it reminded me of Sperner's Lemma in topology, which has a clever odd-even proof. I thought about it for several months, and got the results in the 1968 paper by trying to construct a

dissection into an odd number of triangles of equal areas. They convinced me that there is no such dissection, but I could not prove it.

When I sent the paper to *Mathematics Magazine,* the referee's reaction was predictable. He thought the problem might be fairly easy (although he could not solve it) and was possibly well-known (although he could find no reference to it).

He recommended that I submit the problem to the *Monthly* and, if no one came up with a solution, that the paper should be published. This was done, and the paper appeared three years after I wrote it.

We can introduce a coordinate system such that the four vertices of the square are $(0,0)$, $(1,0)$, $(1,1)$, and $(0,1)$. Thomas [12] proved that this square cannot be cut into an odd number of triangles of equal areas if all the vertices of the triangles have rational coordinates with odd denominators.

Paul Monsky [8] settled the question in 1970 by removing the assumption on the coordinates of the vertices.

But even if Richman had not posed his question, the problem of tiling a polygon by triangles of equal areas would have been raised. In 1984 Victor Klee was writing a paper on convex polytopes. "In order to simplify a computation it was convenient to be able to dissect an arbitrary convex polytope into simplices of equal volumes," he recalled in a letter written in 1990:

> It seemed positively obvious to me that this could be done, so I initially assumed it and proceeded to work out the other details of the proof. Later, when I could not justify the assumption, I had to circumvent it, and found a more complicated argument that did not use the assumption. However, I still felt strongly that a dissection into simplices of equal volume should always be possible. I was familiar with Monsky's theorem. However, for my argument I did not have to restrict the parity of the number of simplices.

In the summer of 1984 at a conference at the University of Oregon I posed to a number of distinguished mathematicians the problem of deciding (for example) whether an arbitrary convex polygon could be dissected into triangles of equal area. Several of them were attracted to the problem, and felt that they were 'on the verge' of the proof that such a dissection was always possible. As far as I remember, none of us seriously suggested that it might not always be possible.

Klee managed to prove his polytope theorem, bypassing his dissection question [6].

In this chapter, we will construct quadrilaterals, even trapezoids, that cannot be cut into triangles of equal areas. Actually, Hales and Strauss in 1982 [2], two years before Klee raised his question, had given a nonconstructive proof that such quadrilaterals exist. So there are, all told, three separate and independent occasions when tiling by triangles of equal areas was considered.

We first develop the algebraic and topological machinery used in Monsky's proof, prove the Richman–Thomas–Monsky theorem and then discuss a variety of related questions and results inspired by that theorem.

1. Algebraic and topological preliminaries

We will make use of one theorem from topology, Sperner's Lemma, which was proved by Sperner in 1928. This theorem was initially used to prove Brouwer's fixed point theorem, which states that any continuous mapping of a closed n-dimensional ball into itself has a fixed point.

We will be using Sperner's Lemma in dimension two. It concerns a simplicial dissection of a plane polygon into triangles. The adjective "simplicial" means that two overlapping (closed) triangles intersect either in a common vertex or in two vertices and the entire edge that joins them. The dissection in Figure 1 is simplicial; the one in Figure 2 is not. For convenience, all the dissections in this chapter will be simplicial. However, Sperner's Lemma and the

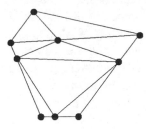

FIGURE 1.   Simplicial.   **FIGURE 2.**   Not simplicial.

theorems in this chapter can be generalized to non-simplicial dissections.

Sperner's Lemma involves the notion of "completeness." A polygon whose vertices are labelled A, B, or C is *complete* if the number of edges whose vertices are labelled A and B is odd. In particular a triangle is complete if and only if its three vertices are labelled A, B, and C. An edge whose vertices are labelled A and B is also called *complete*. The substance of the following exercise will be needed later in the chapter.

Exercise 1.   Introduce n dots on a complete edge E and label each dot A or B. The n dots divide E into $n + 1$ sections.
  (a) Show that the number of sections that have vertices labelled A and B is odd.
  (b) Show that if, on the other hand, the vertices of E are both labelled A or both labelled B, then the number of sections with vertices labelled A and B is even.

Lemma 1.   (Sperner) *Consider a simplicial dissection of a plane polygon. Assume that each vertex is labelled A, B, or C. Then the number of complete triangles is congruent modulo 2 to the number of complete sections on the polygon.*

Proof.   Consider any triangle in the dissection. Place a "pebble" in it next to any edge labelled AB. The procedure is illustrated in Figure 3.

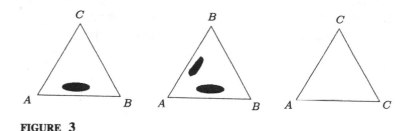

FIGURE 3

A complete triangle collects one pebble. All other triangles collect either two or none. Therefore the total number of pebbles is congruent modulo 2 to the number of complete triangles.

Next count the total number of pebbles in terms of the edges labelled AB. Next to each interior edge AB there are two pebbles. Next to an AB edge on the boundary there is one pebble. Hence the total number of pebbles is congruent modulo 2 to the number of complete edges on the boundary. This establishes Sperner's Lemma.   □

Sperner's Lemma tells us that if the polygon is complete then there is at least one complete triangle in the dissection. For most of our applications this is all we will need from Sperner's Lemma. Later we will state a stronger version, in which orientation plays a role.

Our other tool, valuation, is algebraic [1, pp. 1–15, **13**, pp. 197–202]. A *valuation* on a field F is a function ϕ from F to the real numbers augmented by ∞ that has the following properties:

(1) $\phi(xy) = \phi(x) + \phi(y)$
(2) $\phi(x + y) \geq \min\{\phi(x), \phi(y)\}$
(3) $\phi(x) = \infty$ if and only if $x = 0$.

Note that if ϕ is a valuation on F, then for any positive number k, $k\phi$ is also a valuation. The valuations ϕ and $k\phi$ are called *equivalent*.

Since $\phi(1) = \phi(1 \cdot 1) = \phi(1) + \phi(1)$, it follows that $\phi(1) = 0$. Similarly, $\phi(x) = 0$ for each root of unity and in particular $\phi(-1) =$

0. Thus $\phi(-x) = \phi(-1 \cdot x) = \phi(-1) + \phi(x) = \phi(x)$. Hence $\phi(-x) = \phi(x)$. Also, when x is not 0, $\phi(1/x) = -\phi(x)$. Moreover, if $\phi(x) < \phi(y)$, then a stronger condition than (2) holds, namely $\phi(x + y) = \phi(x)$. To show this, first note that, by property (2), $\phi(x) \leq \phi(x+y)$. But we also have

$$\phi(x) = \phi(x + y - y)$$
$$\geq \min\{\phi(x + y), \phi(-y)\}$$
$$= \min\{\phi(x + y), \phi(y)\}$$
$$\geq \min\{\phi(x), \phi(y)\}$$
$$= \phi(x).$$

Thus $\phi(x) = \min\{\phi(x + y), \phi(y)\}$. Since $\phi(y) > \phi(x)$, we conclude that $\phi(x) = \phi(x + y)$, as claimed.

Exercise 2. Show that if ϕ is a valuation on a field F and x and y, $y \neq 0$, are in F, then $\phi(x/y) = \phi(x) - \phi(y)$.

To construct a valuation on Q, the field of rationals, select a prime p.

Define ϕ_p as follows. First, define $\phi_p(0)$ to be ∞. If $r \in Q$, $r \neq 0$, write r in the form $p^n a/b$, where n, a, and b are integers, and p divides neither a nor b. Define $\phi_p(r)$ to be n, which we can think of as "the number of times that p divides r." The function ϕ_p is called the *p-adic valuation* on Q.

Exercise 3. Prove that ϕ_p is well defined, that is, if $p^m c/d = p^n a/b$, where m, n, a, b, c, and d are integers, and p is relatively prime to $abcd$, then $m = n$.

Exercise 4. Prove that ϕ_p is a valuation on Q.

The next exercise shows that every valuation on Q is of the form $k\phi_p$ for some constant k and prime p.

Exercise 5. Let ϕ be a valuation on Q other than the trivial one that assigns 0 to each number.
(a) Show that for every integer n, $\phi(n) \geq 0$.
(b) Show that there is a prime p such that $\phi(p) > 0$.

(c) Let p' be a prime different from p. Show that $\phi(p') = 0$. Suggestion: There are integers a and b such that $ap + bp' = 1$.

(d) Letting $\phi(p) = k$, show that $\phi = k\phi_p$.

The valuation ϕ_2 is the one we need in the solution of Richman's problem. If we restrict the vertices of the triangles in the dissection to have only rational coordinates, that valuation would suffice. To deal with arbitrary dissections of the square, we need to extend ϕ_2 from Q to a field that includes all the coordinates of the vertices. Fortunately, each p-adic valuation can be extended to all of R, the field of the real numbers (and even to all of C, the field of complex numbers). We pause to sketch how the domain of ϕ_p can be extended beyond Q. The reader may skip over this interlude, for the essential point is that each valuation on Q can be extended in at least one way to any subfield of the reals.

We first sketch briefly how a valuation ϕ_p on Q can be extended to a finite algebraic extension F. Let A be the ring of algebraic integers that lie in F. (A number is an algebraic integer if it is a root of a monic polynomial with integer coefficients.) The algebraic integers form a ring. Each ideal in A (other than 0 and A) is uniquely expressible as a product of prime ideals. In particular, the ideal (p) generated by the prime p has a factorization of the form

$$(p) = P_1^{e(1)} P_2^{e(2)} \cdots P_r^{e(r)},$$

where the $e(i)$'s are positive integers and the P_i's are prime ideals in A. Incidentally, r is not larger than the dimension of F over Q. Pick one of the r prime ideals, say P_1. For x in A, but not 0, define $\phi(x)$ to be $1/e(1)$ times the number of times P_1 appears in the representation of the ideal (x) as a product of prime ideals. Define $\phi(0)$ to be ∞. Note that $\phi(p) = e(1)/e(1) = 1$ and therefore ϕ extends ϕ_p from Q to A. These r extensions are, up to equivalence, the only extensions of ϕ_p to A that have the properties of a valuation.

The extension of ϕ from A to F depends on the fact that every element in F can be expressed as a quotient of elements in A. To see this, let b in F be a root of

$$a_n x^n + a_{n-1} x^{n-1} + \cdots + a_0 = 0,$$

where the coefficients are in Z and a_n is not 0. Then $a_n b$ is a root of

$$a_n \left(\frac{x}{a_n}\right)^n + a_{n-1} \left(\frac{x}{a_n}\right)^{n-1} + \cdots + a_0 = 0$$

or

$$a_n x^n + a_{n-1} a_n x^{n-1} + \cdots + a_0 a_n^n = 0.$$

Cancelling a_n yields a monic polynomial in $Z[x]$ of which the number $a_n b$ is a root. Thus $a_n b$ is an algebraic integer, and $b = (a_n b)/a_n$ is a quotient of elements in A. Define $\phi(b)$ to be $\phi(a_n b) - \phi(a_n)$. Since $\phi(xy)$ is equal to $\phi(x) + \phi(y)$, this extension is single valued.

In particular, if F is a quadratic extension of Q, $F = Q(\sqrt{D})$, there are at most two extensions of ϕ_p to $Q(\sqrt{D})$. For the number of extensions see [3, p. 190]. For instance, for the odd prime p there are two extensions of ϕ_p to $Q(\sqrt{D})$ if and only if p does not divide D and D is a quadratic residue modulo p. For example, there are two extensions of ϕ_7 to $Q(\sqrt{2})$ but only one to $Q(\sqrt{3})$.

In the next few exercises assume that the valuations are defined on all of R.

Exercise 6. Evaluate ϕ_2 at 7, 8, 3/4, and 6.

Exercise 7. Evaluate ϕ_2 at $\sqrt{2}$, $\sqrt[4]{3}$, $3/\sqrt{2}$ and $\sqrt[3]{4}$.

Exercise 8. Evaluate ϕ_2 at $1 + \sqrt{2}$, $\sqrt{2} + \sqrt{5}$, and $2 + \sqrt{1/2}$.

Exercise 9. Find $\phi_2(1 + \sqrt{3})$ as follows. Let $u = 1 + \sqrt{3}$ and $v = 1 - \sqrt{3}$.
(a) Find $\phi_2(uv)$ and $\phi_2(u + v)$.
(b) Show that $\phi_2(u)$ must equal $\phi_2(v)$.
(c) Obtain $\phi_2(u)$.

Exercise 10. There are two possible values for $\phi_2(1 + \sqrt{17})$. Find them.

It is a much simpler matter to extend a valuation ϕ from a field F to the field $F(x)$, where x is transcendental over F. In fact, we may preassign the value of the extension at x to be any real number, say c. It suffices to extend ϕ to $F[x]$. Let ϕ^* be the extension we will construct.

For any monomial ax^n in $F[x]$, $\phi^*(ax^n)$ must be $\phi(a) + nc$. Define ϕ^* on any polynomial as the minimum of its values on any of the monomials in the polynomial. Then extend ϕ^* to $F(x)$ by defining $\phi^*\big(f(x)/g(x)\big) = \phi^*\big(f(x)\big) - \phi^*\big(g(x)\big)$. (We will check it for elements in $F[x]$, leaving the check for elements in $F(x)$ to the reader.) We show that this extension is in fact a valuation.

We first show that

$$\phi^*\big(f(x) + g(x)\big) \geq \min\Big\{\phi^*\big(f(x)\big), \phi^*\big(g(x)\big)\Big\}.$$

Let $f(x) = \sum a_n x^n$ and $g(x) = \sum b_n x^n$. Then

$$\phi^*\big(f(x) + g(x)\big) = \phi^*\left(\sum (a_n + b_n)x^n\right)$$
$$= \min\{\phi(a_n + b_n) + nc\}$$
$$\geq \min\{\min[\phi(a_n), \phi(b_n)] + nc\}$$
$$= \min\{\phi(a_n) + nc, \phi(b_n) + nc\}$$
$$= \min\{\phi^*\big(f(x)\big), \phi^*\big(g(x)\big)\}.$$

In order to deal with $\phi^*\big(f(x)g(x)\big)$ we will need the fact that if $\phi^*\big(f(x)\big) < \phi^*\big(g(x)\big)$, then $\phi^*\big(f(x) + g(x)\big)$ equals $\phi^*\big(f(x)\big)$. To establish this, let m be an index such that $\phi^*(a_m x^m) = \phi^*\big(f(x)\big)$. Then $\phi^*(a_m x^m) < \phi^*(b_m x^m)$ since $\phi^*\big(f(x)\big) < \phi^*\big(g(x)\big)$. Hence $\phi(a_m) < \phi(b_m)$ and therefore $\phi(a_m + b_m) = \phi(a_m)$. Hence

$$\phi^*\big((a_m + b_m)x^m\big) = \phi(a_m + b_m) + mc$$
$$= \phi(a_m) + mc$$
$$= \phi^*\big(f(x)\big).$$

Thus

$$\phi^*\big(f(x) + g(x)\big) = \phi^*\big(f(x)\big).$$

Now we are ready to consider $\phi^*\big(f(x)g(x)\big)$. For each nonnegative integer n we have

$$\phi^*\big(f(x)g(x)\big) \geq \phi^*\left(\left(\sum_{i+j=n} a_i b_j\right) x^n\right)$$

$$= \phi\left(\sum_{i+j=n} a_i b_j\right) + nc$$

$$\geq \phi(a_i b_j) + nc \qquad \text{if} \qquad i+j = n$$

$$= \phi(a_i) + \phi(b_j) + nc$$

$$= \phi(a_i) + ic + \phi(b_j) + jc$$

$$\geq \phi^*\big(f(x)\big) + \phi^*\big(g(x)\big).$$

Next we produce a monomial in $f(x)g(x)$ at which ϕ^* takes on the value $\phi^*\big(f(x)\big) + \phi^*\big(g(x)\big)$.

Let r be the smallest index such that $\phi^*(a_i x^i)$ is equal to $\phi^*\big(f(x)\big)$ and let s be the smallest index such that $\phi^*(b_j x^j)$ is equal to $\phi^*\big(g(x)\big)$. Consider the value of ϕ^* at the monomial in $f(x)g(x)$ of degree $r+s$:

$$\phi^*(a_0 b_{r+s} x^{r+s} + \cdots + a_r b_s x^{r+s} + \cdots + a_{r+s} b_0 x^{r+s}). \qquad (1)$$

If $i+j = r+s$, but i is not r, then $\phi^*(a_i b_j x^{r+s})$ is strictly larger than $\phi^*(a_r b_s x^{r+s})$. Thus (1) is equal to $\phi^*(a_r b_s x^{r+s})$, and therefore equals $\phi^*\big(f(x)\big) + \phi^*\big(g(x)\big)$.

Exercise 11.   Verify that ϕ^* is a valuation on $F(x)$.

We have seen how to extend a valuation from Q to a finite algebraic extension and from any field to an extension by a transcendental element. The extension of a valuation from a field to an infinite algebraic extension is fairly involved, and we refer the reader to the references. In any case, it is possible to extend any of the p-adic valuation on Q to all of the real numbers. Actually, we need to extend it only to fields obtained from Q by an extension by a finite number of elements.

For each valuation ϕ on the field of real numbers we decompose the coordinate plane into three sets, as follows:

$$S_0 = \{(x, y) : \phi(x) > 0 \quad \text{and} \quad \phi(y) > 0\}$$
$$S_1 = \{(x, y) : \phi(x) \le 0 \quad \text{and} \quad \phi(y) \ge \phi(x)\}$$
$$S_2 = \{(x, y) : \phi(x) > \phi(y) \quad \text{and} \quad \phi(y) \le 0\}.$$

Label any point P that is in S_i, P_i. It is easy to see that if $P_i \in S_i$, then the point $P_i - P_0$ is also in S_i, $0 \le i \le 2$.

Exercise 12. Show that $P_i - P_0 \in S_i$.

Exercise 13. Let ϕ be a valuation on R that extends ϕ_2. Determine in which set, S_0, S_1 or S_2, each of these points lies: $(0, 0)$, $(1, 0)$, $(0, 1)$, $(1, 1)$, $(2, 1)$, $(2, 2)$, $(1, 2)$.

The following lemma is the key to analyzing dissections of polygons into triangles of equal areas.

Lemma 2. *Let T be a triangle in the coordinate plane whose vertices are (x_0, y_0), (x_1, y_1), and (x_2, y_2), where (x_i, y_i) is in S_i, relative to a valuation ϕ. Then*

$$\phi(\text{Area of } T) \le -\phi(2).$$

Proof. Since translation by $(-x_0, -y_0)$ does not change areas, and $P_i - P_0 \in S_i, 0 \le i \le 2$, we may assume for convenience that (x_0, y_0) is $(0, 0)$. The area of T is then the absolute value of

$$\frac{1}{2} \begin{vmatrix} x_1 & y_1 \\ x_2 & y_2 \end{vmatrix} = \frac{1}{2}(x_1 y_2 - x_2 y_1).$$

Now, $\phi(x_1) \le 0$ and $\phi(x_1) \le \phi(y_1)$. Also, $\phi(y_2) \le 0$ and $\phi(y_2) < \phi(x_2)$. Thus $\phi(x_1 y_2) < \phi(x_2 y_1)$ and $\phi(x_1 y_2) \le 0$. Hence

$$\phi(\text{Area of } T) = \phi(1/2) + \phi(x_1 y_2) \le \phi(1/2) = -\phi(2). \qquad \square$$

Exercise 14. Use Lemma 2 to show that each line in R^2 meets at most two of the sets S_0, S_1, and S_2.

Exercise 14 shows, for example, that if the vertices of an edge are labelled P_0 and P_1, then any intermediate vertex is labelled P_0 or P_1.

2. The square

All the machinery for analyzing dissections of the square is now in place. But we still need a few definitions. We restrict our attention to simplicial dissections, though the theorems hold for all dissections.

We call a dissection of a polygon into m triangles of equal areas an *m-equidissection*. Let ϕ be a valuation on the reals. A triangle whose vertices are labelled P_0, P_1, and P_2 relative to the valuation ϕ we will call *complete*. (The definitions of "complete," given relative to the letters A, B, and C, carry over to the case when the letters are P_0, P_1, and P_2.) The next lemma is where Sperner's Lemma, valuations, and equidissections come together.

Lemma 3. *Let D be an m-equidissection of a polygon of area A and let ϕ be a valuation on the real numbers. If D contains a complete triangle relative to ϕ, then*

$$\phi(m) \geq \phi(2A).$$

Proof. Let T be a complete triangle in the equidissection. Its area is A/m. By Lemma 2,

$$\phi(A/m) \leq -\phi(2), \quad \text{or} \quad \phi(A) - \phi(m) \leq -\phi(2).$$

Thus

$$\phi(m) \geq \phi(A) + \phi(2) = \phi(2A). \qquad \square$$

We illustrate the utility of Lemma 3 by applying it to equidissections of a square.

Theorem 1. (Richman–Thomas–Monsky) *In any m-equidissection of a square m is even.*

Proof.   Consider the square whose vertices are $(0,0)$, $(1,0)$, $(1,1)$, and $(0,1)$. Relative to a valuation ϕ that extends the 2-adic valuation on Q, these vertices are labelled P_0, P_1, P_1, and P_2, respectively, as shown in Figure 4.

Figure 5 shows a typical simplicial dissection of this square.

By Exercise 14, vertices on the bottom edge of the square can be labelled either P_0 or P_1, on the left edge either P_0 or P_2, on the top edge either P_1 or P_2, and on the right edge either P_1 or P_2. The border is broken into sections by the vertices of the dissection that lie on the border. Therefore the only sections whose ends are

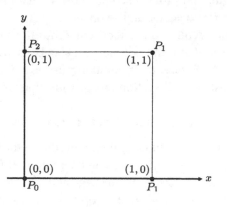

FIGURE 4

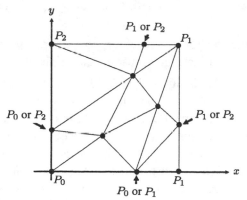

FIGURE 5

labelled P_0 and P_1 are on the bottom edge of the square. There are an odd number of them. By Sperner's Lemma, there is a complete triangle. By Lemma 3

$$\phi(m) \geq \phi\big(2(\text{Area of the square})\big) = \phi(2) = 1.$$

Therefore m is even. $\square$

3. Other polygons

The square can be generalized in many directions: to n-dimensional cubes, to regular polygons, to trapezoids, to quadrilaterals, to centrally symmetric polygons, and so on.

In the first sequel to the theorem about the square, Mead in 1979 [7] proved that if the n-dimensional cube is divided into simplices of equal volumes, their number must be a multiple of $n!$. (It is easy to construct a dissection into any multiple of $n!$ simplices.)

Exercise 15.   Verify the statement in parentheses.

In 1985 Elaine Kasimatis, then a graduate student, was looking for some algebraic topic she could slip into G. D. Chakerian's geometry seminar at Davis. Stein suggested that she report on dissections of the square and cube, a topic that Chakerian grudgingly admitted was geometric. After the talk, Stein asked, "What about the regular pentagon?" Kasimatis's answer was published in 1989 [4]:

> In an equidissection of a regular n-gon, where n is at least 5, the number of triangles must be a multiple of n.

At this point Kasimatis and Stein [5] posed the general question, "If the plane polygon K has an m-equidissection, what can be said about m?" They introduced the *spectrum* of K, defined as the set of integers m such that K has an equidissection into m triangles, and denoted $S(K)$. If m is in $S(K)$ then so is any positive integer multiple of m. If $S(K)$ consists of the multiples of a single integer, m, then K and $S(K)$ are called *principal* and $S(K)$ is also written as $\langle m \rangle$. For instance, the quoted results show that the square, the

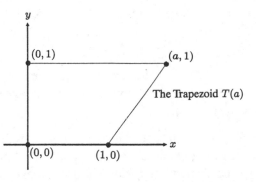

FIGURE 6

n-dimensional cube, and any regular polygon are principal. As we will see, not every polygon is principal and, furthermore, we will construct some polygons that have no equidissections.

Let us begin with trapezoids. Any trapezoid can be carried by an affine mapping into a trapezoid three of whose vertices are $(0,0)$, $(1,0)$, and $(0,1)$. Since an affine transformation magnifies all areas by the same factor, it carries an equidissection into an equidissection. Therefore we consider a trapezoid of the type shown in Figure 6, whose fourth vertex is $(a,1)$.

A trapezoid is affinely equivalent to this trapezoid if the ratio between the lengths of its parallel sides is a. Let $T(a)$ denote the trapezoid in Figure 6.

The next two theorems represent extreme cases, when a is transcendental and when a is rational.

Theorem 2. *If a is transcendental, $T(a)$ has no equidissection.*

Proof. Assume, on the contrary, that $T(a)$ has an m-equidissection. Let p be a prime larger than m. Note that the labelling of the four vertices in Figure 6 is complete relative to any valuation ϕ_p, no matter what $\phi_p(a)$ is. In any dissection of the trapezoid, all the sections whose ends are labelled P_0 and P_1 occur on the bottom edge and there are an odd number of them. Furthermore, the area of $T(a)$ is $(1 + a)/2$.

Applying Sperner's Lemma and Lemma 3, we see that

$$\phi_p(m) \geq \phi_p(a+1).$$

Since $a + 1$ is transcendental, we can prescribe $\phi_p(a + 1)$ to be 1. Thus p divides m, contradicting the choice of p. Hence there is no equidissection of $T(a)$.   □

Exercise 16. A similar theorem does not hold for dissections into quadrilaterals of equal areas, as mentioned in [2]. Show that any polygon can be divided into quadrilaterals of equal areas. (For convenience, restrict the polygons to be convex.)

The next theorem, in contrast with the preceding one, shows that when a is rational, $T(a)$ has equidissections and is principal.

Theorem 3. *Let a be a rational number, $a = r/s$, where r and s are positive, relatively prime integers. Then the spectrum of $T(a)$ is $\langle r+s \rangle$.*

Proof. First of all, $T(a)$ has many $(r + s)$-equidissections. For instance, a diagonal of $T(a)$ cuts $T(a)$ into two triangles whose areas are in the ratio of r to s. Dividing the first triangle into r triangles of equal areas and the second triangle into s triangles of equal areas produces an $(r + s)$-equidissection of $T(a)$. Thus $\langle r + s \rangle$ is a subset of $S(T(a))$.

Next we show that for any m-equidissection, m must be a multiple of $r + s$.

For convenience, consider the affine image of $T(r/s)$ obtained by magnification parallel to the x axis by a factor s. That is, consider the trapezoid T whose vertices are $(0,0)$, $(s,0)$, $(r,1)$, and $(0,1)$, shown in Figure 7.

Let p be a prime dividing $r + s$. We wish to show that for an m-equidissection of T, $\phi_p(m) \geq \phi_p(r + s)$.

Since p divides $r + s$, and r and s are relatively prime, $\phi_p(r) = 0 = \phi_p(s)$. The vertices of T therefore have the complete labelling relative to ϕ_p shown in Figure 8.

As with the square, in any equidissection of T there is a complete triangle. Since the area of T is $(r + s)/2$ we have, by

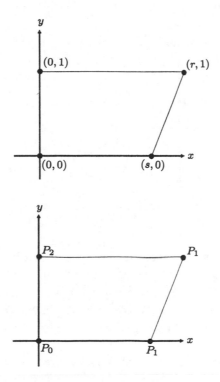

FIGURE 7

FIGURE 8

Lemma 3,

$$\phi_p(m) \geq \phi_p(2(r+s)/2) = \phi_p(r+s).$$

Since $\phi_p(m) \geq \phi_p(r+s)$ for each prime p that divides $r+s$, it follows that $r+s$ divides m.   □

Theorems 2 and 3 raise the question, "What about equidissections of the trapezoid $T(a)$ when a is algebraic but not rational?" The next few theorems sample some of what is known.

Theorem 4.   *Let t_1, t_2, and t_3 be positive integers such that $t_2^2 - 4t_1t_3$ is positive and is not the square of an integer. Let a be one of the roots of the equation $t_3x^2 - t_2x + t_1 = 0$. Then $T(a)$ has an equidissection into $t_1 + t_2 + t_3$ triangles.*

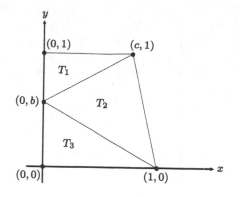

Proof. Let b be a number in the open interval $(0,1)$ and c be a positive number that will be determined later. Cut $T(c)$ into three triangles, T_1, T_2, and T_3, defined as follows. T_1 has vertices $(0,b)$, $(c,1)$, and $(0,1)$; T_2 has vertices $(0,b)$, $(1,0)$, and $(c,1)$; T_3 has vertices $(0,0)$, $(1,0)$, and $(0,b)$. These triangles are shown in Figure 9.

T_1 has area $A_1 = c(1-b)/2$; T_2 has area $A_2 = (cb+1-b)/2$; T_3 has area $A_3 = b/2$. We shall determine b and c so that the areas A_1, A_2, and A_3 are proportional to the integers t_1, t_2, and t_3, respectively. For such a choice of numbers the trapezoid $T(c)$ would then have a $(t_1 + t_2 + t_3)$-equidissection.

The equations

$$\frac{A_2}{A_1} = \frac{t_2}{t_1} \quad \text{and} \quad \frac{A_3}{A_1} = \frac{t_3}{t_1}$$

become

$$\frac{cb+1-b}{c(1-b)} = \frac{t_2}{t_1} \quad \text{and} \quad \frac{b}{c(1-b)} = \frac{t_3}{t_1}.$$

Straightforward algebra shows that $b = ct_3/(1 + ct_3)$, where c is either root of the quadratic $t_3 x^2 - t_2 x + t_1 = 0$. Because $t_2^2 - 4t_1 t_3$ is positive, the root is real. (Note that c is positive and that $0 < b < 1$.)

□

Let us examine the spectrum of $T(a)$ more closely for three values of a that satisfy the conditions of Theorem 4, namely, $a = 2 - \sqrt{2}$, $a = 2 - \sqrt{3}$, and $a = (6 + \sqrt{21})/3$.

In the first case, a is a root of the equation $x^2 - 4x + 2 = 0$. Thus 7 is in $S(T(2 - \sqrt{2}))$. Reasoning similar to that used before shows that for any m-equidissection of the trapezoid $T(2 - \sqrt{2})$,

$$\phi_7(m) \geq \phi_7 \left(2\frac{3 - \sqrt{2}}{2} \right) = \phi_7(3 - \sqrt{2}).$$

Let $d = 3 - \sqrt{2}$ and $e = 3 + \sqrt{2}$. Then $de = 7$ and $d + e = 6$. Hence $\phi_7(d) + \phi_7(e) = 1$ and $\phi_7(d + e) = 0$. If $\phi_7(d)$ were equal to $\phi_7(e)$, we would then have $\phi_7(d) = 1/2 = \phi_7(e)$ and therefore $\phi_7(d+e) \geq 1/2$, contradicting the fact that $\phi_7(d+e) = 0$. Therefore $\phi_7(d)$ and $\phi_7(e)$ are not equal. One of them must therefore be 0 and the other one must be 1. If $\phi_7(3 - \sqrt{2}) = 1$, we already have $\phi_7(m) \geq 1$. If $\phi_7(3 - \sqrt{2}) = 0$, then $\phi_7(3 + \sqrt{2}) = 1$. Let U be the automorphism of $Q(\sqrt{2})$ that takes $\sqrt{2}$ to $-\sqrt{2}$. Define a new valuation ϕ^* on $Q(\sqrt{2})$ by letting $\phi^*(x) = \phi_7(U(x))$. Extend this valuation to R. Then $\phi^*(2 - \sqrt{2}) = 1$. Using this valuation instead of ϕ_7 again shows that 7 divides m. Therefore $T(2-\sqrt{2})$ is principal with spectrum equal to $\langle 7 \rangle$.

If $a = 2-\sqrt{3}$, the argument is slightly different. In this case a is a root of the equation $x^2 - 4x + 1$. Therefore 6 is in the spectrum of $T(2 - \sqrt{3})$. Since the area of this trapezoid is $(3 - \sqrt{3})/2$, $\phi_p(m) \geq \phi_p(3-\sqrt{3})$ for any of its m-equidissections and any prime p. If $p = 3$, we obtain, since $\phi_3(3) = 1$ and $\phi_3(\sqrt{3}) = 1/2$, $\phi_3(3 - \sqrt{3}) = 1/2$. Thus $\phi_3(m) \geq 1/2$, which shows that m is a multiple of 3.

Now consider $\phi_2(3 - \sqrt{3})$. Let $d = 3 - \sqrt{3}$ and $e = 3+\sqrt{3}$. We have $de = 6$ and $d+e = 6$. Hence $\phi_2(d)+\phi_2(e) = 1$ and $\phi_2(d+e) = 1$. If $\phi_2(d)$ and $\phi_2(e)$ were unequal, then one of them would be less than $1/2$ and force $\phi_2(d + e)$ to be less than $1/2$, contradicting the fact that $\phi_2(d + e) = 1$. Therefore $\phi_2(d) = \phi_2(e) = 1/2$. Hence m is a multiple of 2, and the spectrum is $\langle 6 \rangle$.

The case $a = (6+\sqrt{21})/3$ (or $a = (6-\sqrt{21})/3$), however, is not settled. This trapezoid corresponds to $t_1 = 3$, $t_2 = 12$, and $t_3 = 5$.

It has area $(9 + \sqrt{21})/6$ and therefore $\phi_p(m) \geq \phi_p((9 + \sqrt{21})/3)$. We know that $t_1 + t_2 + t_3 = 20$ is in the spectrum.

Letting $d = (9 + \sqrt{21})/3$ and $e = (9 - \sqrt{21})/3$, we have $de = 20/3$ and $d + e = 6$. Reasoning as before, we obtain $\phi_5(d) = 1/2$, from which it follows that 5 divides m. Turning to ϕ_2, we have

$$\phi_2(d) + \phi_2(e) = 2 \qquad \text{and} \qquad \phi_2(d + e) = 1.$$

The assumption that $\phi_2(d)$ is not equal to $\phi_2(e)$ quickly leads to a contradiction. Therefore they are equal and have the value 1. Thus 2 divides m and we can conclude that 10 divides m. All told, we know that

$$\langle 20 \rangle \subset S \left(T \left(\left(6 + \sqrt{21} \right)/3 \right) \right) \subset \langle 10 \rangle.$$

In this case the spectrum is not determined, and we do not know whether it is principal.

Problem 1.   Find the spectrum of the trapezoid $T((6 + \sqrt{21})/3)$.

Problem 2.   Does the trapezoid $T(\sqrt{2})$ have an equidissection? (It can be shown by methods in the next chapter that the answer is no if we demand that the coordinates of the vertices lie in the field $Q(\sqrt{2})$.)

Problem 3.   For which algebraic irrational numbers a does $T(a)$ have an equidissection? (Conjecture: When the real conjugates of a are all positive.)

Problem 4.   Is every trapezoid principal?

There are in fact quadrilaterals that are not principal. For instance, consider the quadrilateral whose vertices are $(0, 0)$, $(1, 0)$, $(0, 1)$, and $(3/2, 3/2)$. Its long diagonal cuts it into two triangles of equal areas. On the other hand, its short diagonal cuts it into a triangle of area $1/2$ and a triangle of area 1, which the long diagonal divides into two triangles of areas $1/2$. Hence both 2 and 3 are in the spectrum of the quadrilateral, and the spectrum is not principal.

This quadrilateral suggests that we consider the class of quadrilaterals that are symmetric with respect to one diagonal.

Let $Q(a)$ be the quadrilateral whose vertices are $(0,0)$, $(1,0)$, (a,a), and $(0,1)$, where a is a number larger than $1/2$ (so that the quadrilateral is convex). The next theorem shows that an infinite number of these quadrilaterals are not principal.

Theorem 5. *Let $a = r/(2s)$, where r and s are relatively prime positive integers, r is odd and is larger than s. Then the spectrum of the convex quadrilateral $Q(a)$ contains 2 and the integer $r + 2sk$ for each nonnegative integer k.*

Proof. Let k be a nonnegative integer and let $x = a/(a+k)$. Divide $Q(a)$ into three triangles with the aid of the vertex $(x, 0)$. (If $k = 0$, there are only two triangles.) The triangles are shown in Figure 10.

The areas of the three triangles are in the proportion

$$x : xa - x + a : a(1 - x),$$

or

$$\frac{a}{a+k} : \frac{a}{a+k}(a-1) + a : a\left(1 - \frac{a}{a+k}\right).$$

Replacing a by $r/(2s)$ and simplifying, we see that the areas are in the proportion

$$s : r - s + sk : sk.$$

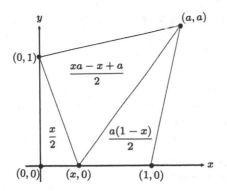

FIGURE 10

Therefore the quadrilateral has an equidissection into $s + (r - s + sk) + sk = r + 2sk$ triangles, as claimed.   □

In the preceding theorem $\phi_2(a) \leq -1$. The next theorem shows that if $\phi_2(a) > -1$, then $Q(a)$ is principal.

Theorem 6. *If $\phi_2(a) > -1$, then the spectrum of $Q(a)$ is $\langle 2 \rangle$.*

Proof. If $-1 < \phi_2(a) \leq 0$, then $Q(a)$ has only one edge labelled $P_0 P_1$, and we can quickly conclude that in any m-equidissection of $Q(a)$ m is even.

If $\phi_2(a) > 0$, apply the shear $(x, y) \to (x/a, y)$, obtaining a quadrilateral with only one edge labelled $P_0 P_1$. The area of this quadrilateral is $a/a = 1$. Working with this quadrilateral, we see again that m is even.   □

Exercise 17.   Fill in the details in the preceding proof.

Theorems 5 and 6 do not cover irrational a for which $\phi_2(a) \leq -1$, for instance, the case $a = \sqrt{3}/2$. So we have a problem similar to the one concerning equidissections of the square.

Problem 5.   Can $Q(\sqrt{3}/2)$ be divided into an odd number of triangles of equal areas?

4. Regular polygons

As already mentioned, Kasimatis proved that the spectrum of the regular n-gon, $n \geq 5$, is $\langle n \rangle$, and therefore is principal. In the general case the proof employs valuations extended to the complex numbers. However, for the regular hexagon it does not, and we present the proof in this case, since it illustrates some of the elements of the general approach.

At first glance we would expect to use the well-known regular hexagon inscribed in the standard unit circle with center at the origin, as shown in Figure 11.

However none of its vertices is labelled P_2 relative to the valuations ϕ_2 or ϕ_3. Consequently we take an affine image of this hexagon. One that works is shown in Figure 12, where three consecutive vertices are $(1, 0)$, $(0, 0)$, and $(0, 1)$, and the other vertices are $(2, 1)$,

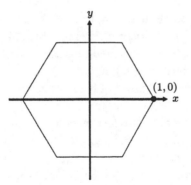

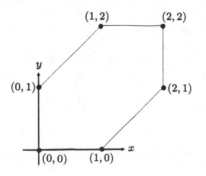

$(2, 2)$, and $(1, 2)$. Of course this step is opposite the more customary approach of trying to analyze a problem in its most symmetric form. In this case we are destroying the rotational symmetry of the hexagon.

The area of the "distorted" polygon shown in Figure 12 is 3. Relative to the valuation ϕ_3 there is only one $P_0 P_1$ edge. Therefore, $\phi_3(m) \geq \phi_3(2 \cdot 3) = 1$ for any m-equidissection of the polygon, and we see that m is a multiple of 3.

The demonstration that m is even is not so direct. Relative to the valuation ϕ_2 the vertices of the distorted hexagon are labelled $P_0, P_1, P_2, P_0, P_1, P_2$, where the vertices are listed counterclockwise starting at the origin. There are two $P_0 P_1$ edges. Note that they are swept out in the same direction, from P_0 to P_1. Fortunately, there is a refined version of Sperner's Lemma that implies that there is a complete triangle in any dissection, from which it follows that 2 divides m. (We state this version below.)

The statement and proof of the stronger Sperner's Lemma are similar to those of the one already discussed.

Consider a simplicial dissection of a polygon in which each vertex is labelled A, B, or C. Let $n_+(ABC)$ be the number of complete triangles in the dissection for which the order A to B to C is counterclockwise. Let $n_-(ABC)$ be the number for which that order is clockwise. Similarly, let $n_+(AB)$ be the number of sections on the polygon labelled AB for which the order A to B gives a counterclockwise orientation to the boundary. Let $n_-(AB)$ be the number for which that orientation is clockwise.

Theorem 7. *Consider a simplicial dissection of a polygon in which each vertex is labelled A, B, or C. Then*

$$n_+(ABC) - n_-(ABC) = n_+(AB) - n_-(AB).$$

The proof is about the same as that of the "non-oriented" Sperner's Lemma. In each triangle place a 1 along an edge AB in which the orientation A to B gives a counterclockwise orientation to the triangle. Place a -1 if the induced orientation is clockwise. Summing up all the 1's and -1's in terms of the triangles and also in terms of the edges immediately yields the theorem.

Exercise 18. Complete the proof of Theorem 7.

5. Centrally symmetric polygons

In 1987 Stein, wondering whether the symmetry of a polygon influences its spectrum, first considered mirror symmetry. Since an isosceles triangle is mirror symmetric and its spectrum is $\langle 1 \rangle$, clearly this type of symmetry puts no constraint on the spectrum. On the other hand, by the theorem on equidissections of a square and by Kasimatis's theorem, the spectrum of a centrally symmetric regular polygon does not contain any odd numbers.

The Richman–Thomas–Monsky theorem implies that in any m-equidissection of a centrally symmetric 4-gon, that is, a parallelogram, m is even. Stein tried to construct a centrally symmetric 6-gon that has an m-equidissection with m odd. The attempt ended with a proof that the spectrum of a centrally symmetric hexagon contains no odd numbers. We present this proof, as simplified by Monsky.

Let the vertices of the hexagon be, in order, D_1, D_2, D_3, D_4, D_5, and D_6. Consider the areas of the six triangles $D_i D_{i+1} D_{i+2}$, where we interpret D_7 as D_1 and D_8 as D_2. Since we are assuming that the hexagon is not a parallelogram, none of these areas is 0. Let us index the vertices so that $\phi_2(\text{Area } D_1 D_2 D_3)$ is the minimum of the six numbers $\phi_2(\text{Area } D_i D_{i+1} D_{i+2})$, $1 \leq i \leq 6$. Exploiting an affine transformation, we make $D_1 = (1, 0)$, $D_2 = (0, 0)$, and $D_3 = (0, 1)$. Let $D_4 = (a, b + 1)$. Then $D_5 = (a + 1, b + 1)$, and $D_6 = (a + 1, b)$, as in Figure 13.

The area of triangle $D_1 D_2 D_3$ is $1/2$. Therefore

$$\phi_2(a/2) = \phi_2(\text{Area } D_2 D_3 D_4) \geq \phi_2(1/2),$$

and we conclude that $\phi_2(a) \geq 0$. Similarly, by considering triangle $D_6 D_1 D_2$, we see that $\phi_2(b) \geq 0$. The labelling of the vertices of the hexagon in Figure 13 relative to the valuation ϕ_2 is almost completely determined, as shown in Figure 14. The main ambiguity is at D_5.

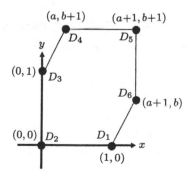

FIGURE 13

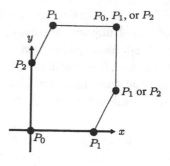

FIGURE 14

If it is not labelled P_0, then there is only one $P_0 P_1$ edge and
Sperner's Lemma applies. If D_5 is labelled P_0, Sperner's Lemma
again applies, though we need the refined version if D_6 is labelled
P_2. Since the area of the hexagon is $1 + a + b$, we then have

$$\phi_2(m) \geq \phi_2\big(2(1 + a + b)\big) \geq \phi_2(2) = 1,$$

showing that in an m-equidissection m is even.

This technique works for a centrally symmetric octagon [11],
but the number of separate cases to be considered grows rapidly
with the number of sides. Using homology groups, Monsky [9] established the theorem in full generality: A bounded centrally symmetric polygon (even with holes) cannot be cut into an odd number
of triangles of equal areas.

Problem 6.   Can a centrally symmetric n-dimensional polyhedron
be cut into an odd number of simplices of equal volumes?

As in the previous two chapters, we find ourselves with more
questions than answers. In those chapters we saw that we do not
fully understand finite cyclic groups, the most elementary groups
imaginable. In this chapter we discovered that even a trapezoid
raises a question that has not been answered. For that reason we
should be grateful that at least some questions have yielded to the
available machinery.

The next chapter is almost the opposite of this one. In it we consider dissections into triangles of certain shapes; their angles rather than their areas will concern us.

References

1. G. Bachman, *Introduction to p-adic numbers and valuation theory,* Academic Press, New York, 1964.

2. A. W. Hales and E. G. Strauss, Projective colorings, *Pacific J. Math.* **99** (1982), 31–43.

3. K. Ireland and M. Rosen, *A classical introduction to modern number theory, Second Ed.,* Springer Verlag Graduate Texts in Mathematics **84**, New York, 1982.

4. E. A. Kasimatis, Dissections of regular polygons into triangles of equal areas, *Discrete and Computational Geometry* **4** (1989), 375–381.

5. E. A. Kasimatis and S. K. Stein, Equidissections of polytopes, *Discrete Math.* **85** (1990), 281–294.

6. V. Klee, Facet-centroids and volume minimization, *Studia Scien. Math. Hung.* **21** (1986), 143–147.

7. D. G. Mead, Dissection of a hypercube into simplices, *Proc. Amer. Math. Soc.* **76** (1979), 302–304.

8. P. Monsky, On dividing a square into triangles, *Amer. Math. Monthly* **77** (1970), 161–164.

9. ——, A conjecture of Stein on plane dissections, *Math. Zeit.* **205** (1990), 583–592.

10. F. Richman and J. Thomas, Problem 5471, *Amer. Math. Monthly* **74** (1967), 329.

11. S. K. Stein, Equidissections of centrally symmetric octagons, *Aequationes Math.* **37** (1989), 313–318.

12. J. Thomas, A dissection problem, *Math. Mag.* **41** (1968), 187–190.

13. B. L. van der Waerden, *Modern Algebra,* Vol. 1, Ungar, New York, 1949.

Chapter 6
Tiling by Similar Triangles

Lajos Pósa wrote his first paper when still in primary school and is well known for his work in graph theory. An enthusiastic teacher, he tries to convey the beauty of mathematics to students of all abilities, from the most talented to the least able, and of all ages, from small children to candidates for the doctorate. Often he holds irregular classes in the most remote towns.

He also organizes summer schools to which he invites students from all over Hungary. In 1987, when preparing for such a session he decided he needed a concrete geometry problem. Now, it is well known [1] that it is possible to cut any polygon into triangles in such a way that the triangles can be assembled to form any preassigned polygon of the same area as that of the original polygon. Pósa wondered whether it is possible to cut an equilateral triangle into 30°-60°-90° triangles that could be put together to form a square. After working on the problem for five minutes he started to like it. After ten minutes he decided that it was interesting enough to assign to his students. After half an hour he grew a little upset, for he still could not solve it. At that point he stopped, for geometry was far from his main interest.

He mentioned his experience to his good friend from school days, Miklós Laczkovich, who found the problem appealing. Laczkovich solved it and went far beyond. In a 25-page paper published in 1990 [2] he proved that it is impossible to cut a square into a finite number of 30°-60°-90° triangles. In fact, he showed, as a

special case of one of his theorems, that it is impossible to cut a
square into a finite number of triangles all of whose angles, when
measured in degrees, are even. As we will see in this chapter, his
proof is almost completely algebraic, involving such tools as fields,
vector spaces, isomorphisms between fields, and complex roots of
unity.

Throughout this chapter the dissections need not be simplicial.

1. The machinery

Let α be the angle determined by lines of slopes m_1 and m_2, neither
of which is parallel to the y axis, as in Figure 1. Let θ_1 and θ_2 be the
corresponding angles of incidence of the two lines, with $\theta_2 > \theta_1$.
Then $\alpha = \theta_2 - \theta_1$ and

$$\tan \alpha = \frac{\tan \theta_2 - \tan \theta_1}{1 + \tan \theta_2 \tan \theta_1} = \frac{m_2 - m_1}{1 + m_1 m_2}.$$

However $\tan \alpha$ is not defined when $\alpha = \pi/2$. Since we want to be
able to work with any angle in a triangle, in particular, with a right
angle, we will use cotangents instead of tangents, and the equation

$$\cot \alpha = \frac{1 + m_1 m_2}{m_2 - m_1}. \tag{1}$$

Rewriting (1) we have

$$(m_2 - m_1) \cot \alpha = 1 + m_1 m_2. \tag{2}$$

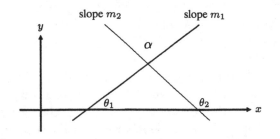

FIGURE 1

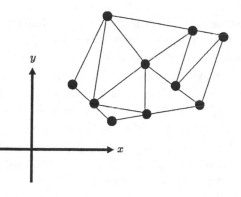

FIGURE 2

So, if we know α and m_2, we can determine m_1, which then lies in the field generated by $\cot \alpha$ and m_2. Therefore, if we are given a sketch of a dissection of a convex polygon P into triangles, as in Figure 2, and the angles of all the triangles and the slopes of the edges of P, we can calculate the slopes of all the edges of the triangles. We could start by determining the slopes of all the edges that meet the boundary, using (2), then gradually work inward from the boundary, repeatedly using (2). Therefore the slopes of all the edges of all the triangles belong to the field generated by the coordinates of the vertices of P and the cotangents of the angles of the triangles.

The following lemma, which will be an important tool, says much more. The assumption in it that P is convex is not necessary, but simplifies the proof a little. By a "vertex of P" we mean a corner vertex, not a point situated in the interior of an edge of P. Thus adjacent edges of P are not parallel and the interior angles of P are less than π.

Lemma 1. *Suppose that a convex polygon P in the xy plane is tiled by a finite set of triangles. Then the coordinates of the vertices of each triangle belong to the field generated by the coordinates of the vertices of P and the cotangents of the angles of the triangles.*

Proof. We may assume that none of the edges of any of the triangles is parallel to the x or y axes. Indeed, if any were parallel to an axis, rotate the polygon about the origin by an angle θ that has rational cosine and sine in such a way that none of the rotated lines is parallel to an axis. (Since there are an infinite number of angles $0 < \theta < 2\pi$ with rational cosine and sine, there is such a rotation.)

Now let $x_1 < x_2 < \cdots < x_r$ be the x coordinates of the vertices of the triangles. We will show that they all lie in the field F generated by the cotangents of the angles of the triangles and the coordinates of the polygon P. (A similar argument goes through for the y coordinates of the vertices.)

First of all, x_1 and x_r are in F since they are the x coordinates of vertices of P. Consider a particular x_i, $2 \leq i \leq r - 1$. There is a number y such that $A = (x_i, y)$ is a vertex of a triangle in the tiling, as shown in Figure 3. Let B and C be the other vertices of one such triangle such that the edge BC crosses the vertical line $x = x_i$ at a point $D = (x_i, u)$. Let α, β, γ, and δ be the angles as labeled in Figure 3. Observe that their cotangents lie in the field F since the slopes of the lines on the three sides of triangle ABC are in F. Let

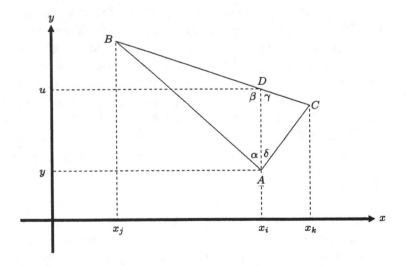

FIGURE 3

B have x coordinate x_j and C, x coordinate x_k. We may assume that $x_j < x_i < x_k$, since the three numbers are distinct.

The length of AD is

$$|u - y| = (x_i - x_j)(\cot \alpha + \cot \beta)$$
$$= (x_k - x_i)(\cot \gamma + \cot \delta). \tag{3}$$

Since $|u - y|$, $x_i - x_j$, and $x_k - x_i$ are positive, there are positive numbers $s = \cot \alpha + \cot \beta$ and $t = \cot \gamma + \cot \delta$ in F such that

$$s(x_i - x_j) = t(x_k - x_i). \tag{4}$$

Now let L be the vector space generated by the numbers x_1, $x_2, \ldots, x_r$ with coefficients in F, that is, all numbers of the form $c_1 x_1 + c_2 x_2 + \cdots + c_r x_r$, c_i in F. Since at least one of the numbers x_1 and x_r is not 0, L contains F. We wish to show that the dimension of L is one, for then L would be just F.

Assume that the dimension of L is greater than one, and let 1 and b be part of a basis. (Since L contains an element of F, we may assume that 1 is a part of a basis for L.) It then follows that b is not in F. For each x in L let $c(x)$ be the coefficient of b in the representation of x as a linear combination of the basis elements: $x = c(x)b + \cdots$.

Then $c(x_i)$ is not 0 for at least one index i. We may assume that $c(x_i)$ is positive for at least one i. (Otherwise replace b in the basis by $-b$.) Let m be the maximum of the numbers $c(x_1), c(x_2), \ldots, c(x_r)$. Let i be the largest of those indices p such that $c(x_p) = m$. Then

$$c\big(s(x_i - x_j)\big) = c\big(t(x_k - x_i)\big).$$

Hence

$$s\big(c(x_i) - c(x_j)\big) = t\big(c(x_k) - c(x_i)\big). \tag{5}$$

By the definition of the index i, $c(x_k) - c(x_i)$ is negative and $c(x_i) - c(x_j)$ is nonnegative. This contradiction shows that the dimension of L over F is one, which completes the proof. $\qquad \square$

Exercise 1. Show that there are an infinite number of angles in the first quadrant whose sine and cosine are rational.

Exercise 2.   Verify equations 4 and 5.

Exercise 3.   Check the claim that for the vertex A there are vertices B and C such that the triangle ABC is in the tiling and the line $x = x_i$ crosses the edge BC.

Exercise 4.   In 1903 Dehn proved that if a square with a rational width is tiled by finitely many rectangles such that the ratio of the width to the length of each rectangle is rational, then the dimensions of each rectangle are rational. Show that this follows immediately from Lemma 1.

The next exercise shows that the lemma could be phrased in terms of slopes rather than cotangents.

Exercise 5.   (Assume Lemma 1.) Suppose that a convex polygon P in the xy plane is tiled by a finite set of triangles, none of whose sides is parallel to the y axis. Then the coordinates of the vertices of each triangle belong to the field generated by the coordinates of the vertices of P and the slopes of the sides of the triangles.

The next lemma involves the notions of an oriented triangle and the relation between the cotangent of an angle and the slopes of its two arms. We pause to describe these ideas.

Let $A = (a_1, a_2)$, $B = (b_1, b_2)$, and $C = (c_1, c_2)$ be three non-collinear points in the xy plane. The triangle with vertices A, B, and C is assigned an orientation by ordering its three vertices in a particular sequence. For instance, in Figure 4 the order "first A, then

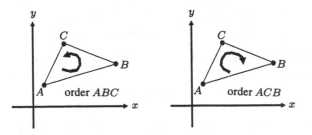

FIGURE 4

B, then C," which we abbreviate to A, B, C, is counterclockwise, but the order A, C, B, for instance, is clockwise.

The notion of orientation can also be expressed in terms of the sign of a determinant. Consider the case $A = (0, 0)$, $B = (1, 0)$, and $C = (0, 1)$. The order A, B, C is counterclockwise and the 2 by 2 determinant

$$\begin{vmatrix} B - A \\ C - A \end{vmatrix} = \begin{vmatrix} (1, 0) - (0, 0) \\ (0, 1) - (0, 0) \end{vmatrix} = \begin{vmatrix} 1 & 0 \\ 0 & 1 \end{vmatrix} = 1$$

is positive. More generally, if we are given the coordinates of three noncollinear vertices, A, B, and C, we can determine whether the order A, B, C is counterclockwise or clockwise by computing the determinant

$$\begin{vmatrix} B - A \\ C - A \end{vmatrix}. \tag{6}$$

When this determinant is positive, the order is counterclockwise; when it is negative, the order is clockwise.

Exercise 6. Is the order $(2, 3)$, $(1, 4)$, $(3, 5)$ clockwise or counterclockwise? Solve by drawing the points and also by using the determinant.

Exercise 7. Show by using vectors or elementary geometry that the absolute value of (6) is twice the area of the triangle whose vertices are A, B, and C.

We also need to express the cotangents of the angles of any triangle ABC in terms of the coordinates of its vertices.

Assume that the order A, B, C is counterclockwise and let α be the angle at the vertex A, as in Figure 5.

The slope of AC is

$$\frac{c_2 - a_2}{c_1 - a_1}$$

and the slope of AB is

$$\frac{b_2 - a_2}{b_1 - a_1}.$$

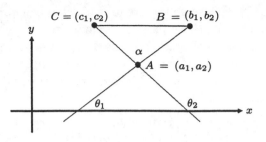

FIGURE 5

Substituting these values in (1) gives

$$\cot \alpha = \frac{(c_2 - a_2)(b_2 - a_2) + (c_1 - a_1)(b_1 - a_1)}{(c_2 - a_2)(b_1 - a_1) - (c_1 - a_1)(b_2 - a_2)}. \qquad (7)$$

Note that the denominator, which is the determinant

$$\begin{vmatrix} B - A \\ C - A \end{vmatrix},$$

is positive.

If, on the other hand, the order A, B, C is clockwise, the expression for $\cot \alpha$ changes. In this case the order A, C, B is counterclockwise and the formula for $\cot \alpha$ is obtained from (7) by switching the roles of B and C. The numerator in (7) remains the same, but the denominator changes sign. This brings us to a key lemma.

Lemma 2. *Let A, B, and C be noncollinear points and let F be a subfield of R that contains their coordinates. Let $\phi : F \to R$ be an isomorphism from F onto a subfield of R. Define the mapping $\Phi : F \times F \to R \times R$ by*

$$\Phi(x, y) = (\phi(x), \ \phi(y)).$$

Let $A' = \Phi(A)$, $B' = \Phi(B)$, and $C' = \Phi(C)$. Let α be the angle at A in triangle ABC and α' be the angle at A' in triangle $A'B'C'$. Then

$$\cot \alpha' = \phi(\cot \alpha)$$

if the orientations A, B, C and A', B', C' are the same, and

$$\cot \alpha' = -\phi(\cot \alpha)$$

if they are different.

Exercise 8. Prove that the points A', B', and C' in Lemma 2 are not collinear.

Exercise 9. Prove Lemma 2 in the case that the orientation A, B, C is counterclockwise and the orientation A', B', C' is clockwise.

Exercise 10. Prove that Φ takes collinear points to collinear points.

Keep in mind that Φ is defined only on $F \times F$, not necessarily on the whole xy plane. It need not be continuous. In fact, it might not even preserve "betweeness." For instance, let $F = Q(\sqrt{2})$ and $\phi : F \to R$ be defined by $\phi(r_1 + r_2\sqrt{2}) = r_1 - r_2\sqrt{2}$, where r_1 and r_2 are rational numbers. The point $(1, 0)$ is between the points $(0, 0)$ and $(\sqrt{2}, 0)$, but $\Phi(1, 0)$ is not between $\Phi(0, 0)$ and $\Phi(\sqrt{2}, 0)$, as may easily be checked.

Exercise 11. Show that the isomorphism $\phi : Q(\sqrt{2}) \to R$, defined above, is not continuous.

Now, if α, β, and γ are the three angles in a triangle ABC, then at least two of them are acute and therefore have positive cotangents. If the mapping Φ in Lemma 2 preserves the orientation of the triangle, that is, the orientation $\Phi(A)$, $\Phi(B)$, $\Phi(C)$ is the same as the orientation A, B, C, then at least two of the numbers $\phi(\cot \alpha)$, $\phi(\cot \beta)$, and $\phi(\cot \gamma)$ are positive, since they are the cotangents of the three angles in a triangle.

With these tools at our disposal, we are ready to analyze dissections of a convex polygon P into triangles.

Let the vertices of P be $V_1, V_2, \ldots, V_n$ in counterclockwise order. We represent the boundary of P by the formal sum

$$V_1 V_2 + V_2 V_3 + \cdots + V_n V_1.$$

We call such a formal sum a chain. (Appendix C gives a precise definition of a formal sum.) The ordered pair $V_{i-1} V_i$ may be thought

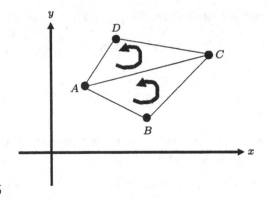

FIGURE 6

of as a vector from V_{i-1} to V_i. If P and Q are two points in the xy plane, we regard the ordered pair QP as the negative of the ordered pair PQ. In other words, $PQ + QP = 0$. Also, if an edge AB is divided by an intermediate point C into two edges AC and CB, we shall write $AC + CB = AB$. The boundary of triangle ABC with orientation A, B, C will be written $AB + BC + CA$. As a result, if two nonoverlapping triangles share a common edge and both have a counterclockwise orientation, their common edges "cancel" algebraically. For instance, consider triangles ABC and ACD in Figure 6, both with a counterclockwise orientation. The formal sum of their algebraic boundaries is

$$(AB + BC + CA) + (AC + CD + DA),$$

which reduces to

$$AB + BC + CD + DA,$$

the formal boundary of their union. This cancellation property extends to any family of similarly oriented triangles that tiles a polygon: the sum of their boundaries is the boundary of the polygon. The sum of the boundaries of the five triangles in Figure 7 is

$$(V_4V_1 + V_1B + BV_4) + (V_1C + CB + BV_1) + (BC + CV_2 + V_2B)$$
$$+(AV_2 + V_2V_3 + V_3A) + (AV_3 + V_3V_4 + V_4A),$$

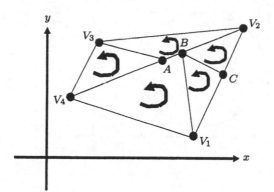

FIGURE 7

which reduces, due to inner cancellations, to

$$V_1C + CV_2 + V_2V_3 + V_3V_4 + V_4V_1,$$

hence to

$$V_1V_2 + V_2V_3 + V_3V_4 + V_4V_1,$$

the counterclockwise boundary of the polygon in Figure 7.

The proof of the next lemma uses another way of recording the orientation of a polygon. If Γ is the boundary of a convex polygon, the value of the line integral $\oint_\Gamma x\,dy$ depends on whether we sweep out Γ in a counterclockwise or clockwise manner. Reversing the orientation switches the sign of the integral.

Consider the case when Γ is oriented counterclockwise, as in Figure 8. Break Γ into a "right-hand" path Γ_2 and a "left-hand" path Γ_1. On Γ_2 y increases, and on Γ_1 y decreases. We have

$$\oint_\Gamma x\,dy = \oint_{\Gamma_2} x\,dy + \oint_{\Gamma_1} x\,dy = \int_a^b x_2\,dy - \int_a^b x_1\,dy = \int_a^b (x_2 - x_1)\,dy,$$

which is the area of the region that Γ bounds. If Γ were clockwise, then $\oint_\Gamma x\,dy$ would be the negative of the area that Γ bounds.

Lemma 3. *Let P be a convex polygon with vertices $V_1, V_2, \ldots, V_n$, listed counterclockwise. Suppose that P is tiled by the triangles T_1,*

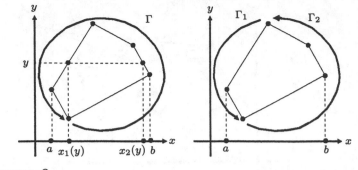

FIGURE 8

$T_2, \ldots, T_m$. *Let the vertices of T_j be A_j, B_j, and C_j, listed counterclockwise, $1 \leq j \leq m$. Let F be a real field that contains the coordinates of all the vertices of the triangles and $\phi : F \to R$ an isomorphism that leaves the coordinates of the vertices of P fixed. Let Φ be as in Lemma 2. Then there is at least one triangle T_j such that the order $\Phi(A_j)$, $\Phi(B_j)$, $\Phi(C_j)$ is counterclockwise.*

Proof.   The algebraic boundary of T_j is $A_jB_j + B_jC_j + C_jA_j$. The sum of the boundaries for all the triangles T_j, $1 \leq j \leq m$, is the algebraic boundary of P, $V_1V_2 + V_2V_3 + \cdots + V_nV_1$. Thus the sum of the chains

$$\delta_j = \Phi(A_j)\Phi(B_j) + \Phi(B_j)\Phi(C_j) + \Phi(C_j)\Phi(A_j), \qquad 1 \leq j \leq m,$$

is the chain

$$\delta = V_1V_2 + V_2V_3 + \cdots + V_nV_1,$$

since $\Phi(V_i) = V_i$, $1 \leq i \leq n$. Therefore

$$\sum_{j=1}^{m} \oint_{\delta_j} x\,dy = \oint_{\delta} x\,dy. \qquad (8)$$

Since the right-hand side of (8) is positive, at least one summand on the left is positive. Let j be an index such that $\oint_{\delta_j} x\,dy$ is positive. Then the orientation $\Phi(A_j)$, $\Phi(B_j)$, $\Phi(C_j)$ is counterclock-wise. This completes the proof.   □

This lemma is the basis of the next one, which concerns angles rather than orientations.

Lemma 4. *Let P be a convex polygon with vertices $V_1, V_2, \ldots, V_n$. Suppose that P is tiled by the triangles $T_1, T_2, \ldots, T_m$. Let F be a real field that contains the coordinates of all the vertices of P and the cotangents of all the angles of the triangles. Let $\phi : F \to R$ be an isomorphism that leaves the coordinates of all the V_i fixed. Then there is an integer j such that at least two of the numbers $\phi(\cot \alpha_j)$, $\phi(\cot \beta_j)$, and $\phi(\cot \gamma_j)$ are positive, where α_j, β_j, and γ_j are the angles of T_j.*

Proof.   By Lemma 1, F contains the coordinates of the vertices of all the triangles. Let Φ be as in Lemma 2. By Lemma 3, there is an integer j such that Φ preserves the orientation of T_j, that is, if T_j is given the counterclockwise orientation A_j, B_j, C_j, then the orientation $\Phi(A_j)$, $\Phi(B_j)$, $\Phi(C_j)$ is also counterclockwise. Now let the angles of T_j be α_j, β_j, and γ_j. By Lemma 2, $\phi(\cot \alpha_j)$, $\phi(\cot \beta_j)$, and $\phi(\cot \gamma_j)$ are the cotangents of the angles in the triangle whose vertices are $\Phi(A_j)$, $\Phi(B_j)$, and $\Phi(C_j)$. Since every triangle has at least two acute angles, hence at least two angles with positive cotangents, the lemma follows.   □

2. Applications

We are now in position to answer Pósa's question.

Theorem 1.   *It is impossible to tile a square with a finite number of $30° - 60° - 90°$ triangles.*

Proof.   We assume, without loss of generality, that the vertices of the square are $(0,0)$, $(1,0)$, $(1,1)$, and $(0,1)$. Let F be $Q(\sqrt{3})$, a field that contains the coordinates of the vertices of the square and also the cotangents of the angles of the triangles in the tiling, since $\cot 30° = \sqrt{3}$, $\cot 60° = 1/\sqrt{3}$, and $\cot 90° = 0$.

Define $\phi : F \to R$ by setting $\phi(r_1 + r_2\sqrt{3}) = r_1 - r_2\sqrt{3}$, where r_1 and r_2 are rational numbers. The values of ϕ at the cotangents of the three angles of every triangle are $-\sqrt{3}$, $-1/\sqrt{3}$, and 0. This contradicts Lemma 4 and therefore proves the theorem. $\square$

Exercise 12.

(a) Use Figure 9 to show that the square can be tiled by an infinite number of triangles similar to any given right triangle.

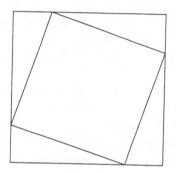

FIGURE 9

(b) Where in the proof of Theorem 1 is the assumption that there are only a finite number of triangles used?

Exercise 13. Outline the flow of the key arguments that lead up to the proof of Theorem 1.

Exercise 14. (Contributed by Mark Chrisman.) The following steps outline an elementary proof that there is no simplicial dissection of a square into 30°-60°-90° triangles.

(a) Assume that there is such a dissection and that the length of a side of the square is 1. Let the length of the shortest side of all the triangles be s. Show that the length of the shortest side in each of the triangles is of the form $2^m 3^{n/2} s$, where m and n are integers.

(b) Obtain an equation for s by using the fact that the sum of the areas of the triangles is the area of the square.

(c) Obtain another equation for s by using the fact that an edge of the square is the union of edges of the triangles.

(d) Using (b) and (c), obtain a contradiction.

(e) What properties of the square were used in this argument? To what polygons does the argument apply?

(f) What properties of the 30°-60°-90° triangle were used in this argument? To what triangles does the argument apply?

To treat the more general case than that covered in Theorem 1, where we now assume only that all the angles of all the triangles in the dissection have an even number of degrees, we have to examine the effect of an isomorphism on $\cot\theta$.

Recall that

$$\cos\theta = \frac{e^{i\theta} + e^{-i\theta}}{2} \quad \text{and} \quad \sin\theta = \frac{e^{i\theta} - e^{-i\theta}}{2i}.$$

Thus

$$\cot\theta = i\frac{e^{i\theta} + e^{-i\theta}}{e^{i\theta} - e^{-i\theta}}.$$

Let ω be the complex number of angle $2\pi/4n = \pi/2n$, where n is a positive integer, and of modulus 1. Thus $\omega = e^{\pi i/2n}$. Note that ω is a primitive $(4n)$th root of unity and that $i = \omega^n$. If the integer a is not a multiple of n, then $\cot(a\pi/n)$ is defined and we have

$$\cot\frac{a\pi}{n} = \omega^n\frac{\omega^{2a} + \omega^{-2a}}{\omega^{2a} - \omega^{-2a}}, \tag{9}$$

which therefore lies in the field $Q(\omega)$.

Lemma 5. *Let n be a positive integer and $\omega = e^{\pi i/2n}$. Let k be an odd integer relatively prime to n. Then there is an automorphism $\phi :$ $Q(\omega) \to Q(\omega)$ such that*

$$\phi\left(\cot\frac{a\pi}{n}\right) = (-1)^{(k-1)/2}\cot\frac{ak\pi}{n}$$

for every integer a not divisible by n.

Proof. Since $(k, 4n) = 1$, the number ω^k is a primitive $(4n)$th root of unity and there is therefore an automorphism $\phi : Q(\omega) \to Q(\omega)$

such that $\phi(\omega) = \omega^k$. Consequently, in view of (9),

$$\phi(\cot \frac{a\pi}{n}) = \omega^{kn} \frac{\omega^{2ak} + \omega^{-2ak}}{\omega^{2ak} - \omega^{-2ak}}$$

$$= \omega^{(k-1)n} \omega^n \frac{\omega^{2ak} + \omega^{-2ak}}{\omega^{2ak} - \omega^{-2ak}}$$

$$= (\omega^{2n})^{(k-1)/2} \cot \frac{ak\pi}{n}$$

$$= (-1)^{(k-1)/2} \cot \frac{ak\pi}{n}. \qquad \square$$

Theorem 2. *Suppose that each vertex of the convex polygon P has only rational coordinates. Assume that P is tiled by a finite number of triangles whose angles are rational multiples of π, hence of the form $a_i\pi/n$ for some fixed integer n and integers a_i. Then n is a multiple of 4.*

Proof. Let F be the field generated by the cotangents of all the angles of the triangles in the tiling, that is, by the numbers $\cot(a_i\pi/n)$. As we observed, all these numbers lie in $Q(\omega)$, where ω is a primitive $(4n)$th root of unity.

Suppose that n is not a multiple of 4. Let $k = 2n + 1$ if n is odd and let $k = n + 1$ if n is even (hence $n \equiv 2 \pmod 4$). Since k is odd and relatively prime to n, we have $(k, 4n) = 1$. The definition of k assures us also that $(k - 1)/2$ is an odd integer and that $k \equiv 1 \pmod n$.

By Lemma 5 there is an automorphism $\phi : Q(\omega) \rightarrow Q(\omega)$ such that

$$\phi\left(\cot \frac{a_i\pi}{n}\right) = (-1)^{(k-1)/2} \cot \frac{a_i k\pi}{n} \qquad (10)$$

for each a_i. The restriction of ϕ to F is then an isomorphism of F into R.

Now, since $(k - 1)/2$ is odd and $k \equiv 1 \pmod n$, (10) reduces to

$$\phi\left(\cot \frac{a_i\pi}{n}\right) = -\cot \frac{a_i\pi}{n}.$$

Thus if θ is an angle in any of the triangles of the alleged tiling, $\phi(\cot\theta) = -\cot\theta$. If α, β, and γ are the angles in any of these triangles, at least two of the numbers $\phi(\cot\alpha)$, $\phi(\cot\beta)$, and $\phi(\cot\gamma)$ are negative. This contradiction of Lemma 4 completes the proof.

$\square$

Exercise 15.   Show that Theorem 2 implies that it is impossible to tile a square with triangles all of whose angles, when measured in degrees, are even integers.

In a letter Laczkovich wrote in 1992, he remarked, "Next I wanted to determine all the triangles that tile a square. For a long time I was convinced that only right triangles can tile a square or even a rectangle. When I discovered [Figure 10] quite accidentally, I was shocked." A polygon P is said to tile a polygon Q if Q can be tiled by polygons similar to P. The angles in each triangle in Figure 10 are $\pi/6$, $\pi/6$, and $2\pi/3$.

Exercise 16.
(a) Verify that Figure 10 is indeed a tiling of the 1 by $\sqrt{3}$ rectangle.
(b) Prove that the $\pi/6$, $\pi/6$, $2\pi/3$ triangle does not tile a square.

Laczkovich went on to show that there are only three triangles that are not right triangles that tile the square. Their angles are $(\pi/8, \pi/4, 5\pi/8)$, $(\pi/4, \pi/3, 5\pi/12)$, and $(\pi/12, \pi/4, 2\pi/3)$. These

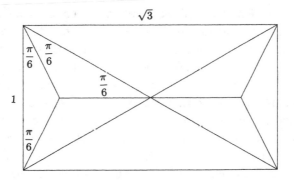

FIGURE 10

and the triangle used in Figure 10 are the only triangles other than right triangles that tile some rectangle.

Exercise 17.   Show that three 15°-75°-90° triangles can tile a 1 by 4 rectangle, hence twelve of them can tile a square.

Laczkovich discovered a good deal more about tilings by similar triangles, and his results presented in this chapter may tempt the reader to look at his paper. Since it appeared, he has also considered the question, "Which pairs of triangles T_1 and T_2 have the property that similar copies of T_1 tile T_2." Working with Szekeres, he also investigated tilings of a square by a finite number of rectangles similar to a given rectangle R. To describe one of their results, let the ratio of the sides of R be r. They proved that a square can be tiled by a finite number of rectangles similar to R if and only if r is an algebraic number such that the real part of each of its conjugates is positive. Using this result, they then produced a right triangle with legs a and b such that similar copies of it tile the square, but rectangles similar to the one with sides a and b do not. (To be specific, b/a is the real root of $x^3 + x - 1 = 0$.)

3. Retrospective

With this chapter we complete our sampling of the algebra of tiling. (The next chapter is the proof of Rédei's Theorem.)

As we look back over these chapters, we can see how the interplay of question and answer, followed by new questions, breathes life into mathematics. In our case the questions came from within mathematics, in particular, from geometry. But they can come from any discipline, such as physics, economics, or computer science.

Minkowski's question about cube tilings led to a question about finite abelian groups. Hajós's answer, in turn, stimulated Rédei to ask, "Must the sets in the factorization be cyclic, or is there a more general theorem, where the sets are arbitrary but of prime orders?" His theorem then gave us an insight into the tilings of space by clusters that consist of a prime number of cubes.

The question, "Which semicrosses or crosses tile space?," quickly suggested looking at splittings of finite abelian groups. These splittings raised questions about exact sequences and cyclic groups, especially groups of prime orders. We were left with more open questions than faced us when we started.

Packing and covering by the semicross and cross raised different questions, including further questions about finite abelian groups, especially cyclic groups. The presence of such questions reminds us that even though we have a structure theorem for finite abelian groups and know that each one is a product of cyclic groups of prime power orders, they still keep many of their secrets.

In spite of a variety of results about cutting polygons into triangles of equal areas, we were left with unanswered questions as primitive as the ones we started with. For instance, we do not know whether the quadrilateral with vertices $(0, 0)$, $(1, 0)$, $(0, 1)$, $(\sqrt{3}/2, \sqrt{3}/2)$ can be cut into an odd number of triangles of equal areas.

Tiling a polygon by triangles with prescribed angles led to algebraic questions completely different from those in the earlier chapters. Though we met some surprising tilings, we are left with the general question of when similar copies of one polygon tile another polygon.

Clearly, we do not need to venture into outer space or the almost inaccessible regions of the Amazon forest to enter territories that have never been explored. Mathematics still offers a cornucopia of challenges to even the most daring spirits, to anyone who is compelled to enter worlds where no one has ever trod.

References

1. H. Eves, *A Survey of Geometry*, Allyn and Bacon, 1972.
2. M. Laczkovich, Tilings of polygons with similar triangles, *Combinatorica* **10** (1990), 281–306.

Chapter 7
Rédei's Theorem

As a young mathematician, G. Hajós prepared a Ph.D. thesis on certain determinant identities. The chairman of his doctoral committee, L. Fejér, whose name is closely associated with Fourier analysis, feeling that the result did not match the outstanding talent of the candidate, rejected the thesis. This is why Hajós turned to Minkowski's famous unsolved conjecture.

In 1938 Hajós formulated the problem in terms of factorizations of groups and, making use of this reformulation, refuted Furtwängler's conjecture about multiple cube tilings, described in Chapter 1. This time his thesis met Fejér's legendary high standards.

Almost everyone, on first meeting the group theoretical equivalent of Minkowski's conjecture, tends to think that the solution of the problem should be immediate. So did Hajós. However, it took him three years to settle the conjecture. Looking back years later, he said that the problem had been extremely deceiving. It had offered many ways of attack but all but one led nowhere. Thinking about the problem almost constantly, he was able to pose it in many different versions. As he said, "When I had to walk up to the 5th floor I might make up my mind to find a new version on the way."

Eventually he succeeded, obtaining his beautiful proof in 1941. It is so algebraic that there is no discernible connection between its lemmas and the geometry of the conjecture. This was why Hajós was reluctant to talk about his proof. "Yes, I had a proof" he used to say, "but I couldn't see what was really happening."

It seems that he eventually found a more satisfactory view of his proof, since some 30 years later, in 1970, he announced a seminar on Minkowski's conjecture at Eötvös University in Budapest. Unfortunately, Hajós died, at the age of fifty nine, in 1971 before the seminar was to be held.

Hajós's proof is based on a deep investigation of the zero divisors in a group ring $Z(G)$. L. Rédei wanted to prove the theorem using only group theoretical concepts, without referring to the group ring, since the theorem itself is formulated in terms of groups. After an extensive analysis he eventually found its group theoretical core and discovered another approach to the problem: the technique of replacing factors in a factorization. His proof, which does not rely on the group ring, led to a broad generalization of Hajós's theorem. This breakthrough, accomplished in 1965, is ample reward for his 25-year-long search.

Rédei [3] proved the following theorem.

Theorem 1. *Let G be a finite abelian group and $A_1, A_2, \ldots, A_n$ subsets of G such that each contains the identity element, each has a prime number of elements, and $G = A_1 A_2 \cdots A_n$ is a factorization of G. Then at least one of the factors A_i is a subgroup of G.*

Since Hajós's theorem is equivalent to its special case when the cyclic factors are of prime cardinalities, Rédei's theorem indeed generalizes the Hajós–Minkowski theorem.

The proof we present in this chapter [1] is a much simplified version of Rédei's proof. Though he did not make use of group rings, we will, but only briefly.

When each factor in a factorization contains the identity element, we call the factorization *normalized*. If we omitted this condition, then the theorem would state that at least one of the factors A_i is a coset of a subgroup of G.

The argument begins by establishing Rédei's theorem in the special case that G is a p-group and the factors are cyclic subsets. (This is a special case of Hajós's theorem.)

Then we establish Rédei's theorem for cyclic p-groups, using the irreducibility (over the field of rationals) of the cyclotomic polynomials whose roots are the primitive (p^n)th roots of unity.

A much more involved proof then treats the noncyclic group of order p^2. A short induction then establishes Rédei's theorem for any finite abelian p-group.

Finally, another induction yields the proof for the general case of any finite abelian group.

Critical to the arguments are replacement principles which permit us, under certain conditions, to replace a factor in any factorization of a group by a related subset of G. As we proceed we will develop these principles, all of which are special cases of one general principle.

Exercises 1 and 2 show that the condition that the factors are of prime cardinalities cannot be removed from Rédei's hypotheses.

Exercise 1.   Let G be the cyclic group of order 8 with generator u.
(a) Show that neither $A = \{e, u, u^4, u^5\}$ nor $B = \{e, u^2\}$ is a subgroup of G.
(b) Verify that $G = AB$ is a factorization of G.

Exercise 2.   Let G be the direct product of cyclic groups of order 8 and 2 and let u and v a basis for G such that $u^8 = v^2 = e$.
(a) Show that neither $A = \{e, u, v, uv\}$ nor $B = \{e, u^2, u^4, u^6v\}$ is a subgroup of G.
(b) Verify that $G = AB$ is a factorization of G.

1. Hajós's theorem for p-groups

Our proof of Rédei's theorem requires a special case of Hajós's theorem, namely the case when G is a p-group. The proof depends on a theorem that allows us to replace a factor in a factorization by another subset.

If A and A' are subsets of a finite abelian group G such that for every subset B of G, if $G = AB$ is a factorization, then $G = A'B$ is also a factorization, we say that A is *replaceable* by A'. In this section we need only a fairly simple replacement principle. Later we develop more general replacement principles. .

Lemma 1. (First replacement principle) *Let $G = AB$ be a normalized factorization of the finite abelian group G. Suppose that $A = \{e, a, a^2, \ldots, a^{p-1}\}$, where p is a prime. Let k be an integer relatively prime to p and $A' = \{e, a^k, a^{2k}, \ldots, a^{(p-1)k}\}$. Then $G = A'B$ is also a factorization of G.*

Proof. Since $G = AB$ is a factorization, the sets

$$eB = B, aB, a^2B, \ldots, a^{p-1}B \tag{1}$$

form a partition of G. Thus aG, which is G again, is partitioned by the sets

$$aB, a^2B, \ldots, a^pB,$$

from which we conclude that $a^pB = B$. This implies that for any integer m (negative or positive), a^mB equals a^jB, where $0 \le j \le p - 1$, $j \equiv m \pmod{p}$.

Now consider $A'B$, which is the union of the sets

$$eB = B, a^kB, a^{2k}B, \ldots, a^{(p-1)k}B. \tag{2}$$

Since k and p are relatively prime the sets (1) are a permutation of the sets (2). Thus $G = A'B$ is a factorization, proving the lemma. $\square$

Exercise 3. Show that if k and p are not relatively prime then A cannot necessarily be replaced by A'. (Hint: Use a factorization of a group of order 4.)

Let $G = A_1 \cdots A_n$ be a factorization of the finite abelian p-group G, where $A_i = \{e, a_i, a_i^2, \ldots, a_i^{p-1}\}$. Any s of the elements $a_1, \ldots, a_n$ generate a subgroup of order at least p^s and the n elements $a_1, a_2, \ldots, a_n$ generate a subgroup of order exactly p^n, that is, G itself.

Exercise 4. Verify the preceding observations.

We introduce some notations. Let g be an element of the finite abelian group G. The order of g will be denoted by $|g|$ and the subgroup generated by g will be denoted $\langle g \rangle$. Let A be a subset of G. The cardinality of A and the subgroup spanned by A will be denoted by $|A|$ and $\langle A \rangle$ respectively. Let $A = \{a_1, \ldots, a_n\}$. If $G = A_1 \cdots A_n$, where $A_i = \{e, a_i, a_i^2, \ldots, a_i^{p-1}\}$, then $|\langle B \rangle| \geq p^{|B|}$ for each $B \subset A$. This property of the factorization has a remarkable consequence described in Lemma 2, which is the key to Hajós's theorem for p-groups. Its proof makes use of the next exercise.

Exercise 5. Let G be an abelian group and H a proper subgroup of G. Assume that there are factorizations $H = C_1 C_2 \cdots C_r$ and $G/H = D_1 D_2 \cdots D_s$. For each i, $1 \leq i \leq s$ let D_i^* be a set of representatives of the cosets of H in D_i. Show that

$$G = C_1 C_2 \cdots C_r D_1^* D_2^* \cdots D_s^*$$

is a factorization of G.

Lemma 2. *Let A be a subset of an abelian p-group G such that $|\langle A \rangle| = p^{|A|}$ and $|\langle B \rangle| \geq p^{|B|}$, for all $B \subset A$. Then for each $a \in A$ there exists a power of p, say $s(a)$, such that*

$$\langle A \rangle = \prod_{a \in A} \{e, a^{s(a)}, a^{2s(a)}, \ldots, a^{(p-1)s(a)}\}$$

is a factorization of $\langle A \rangle$ and at least one of the factors is a subgroup of G.

Proof. The lemma holds when $|A| = 1$. This suggests an induction on $n = |A|$. For a given n, $h(A) = \prod_{a \in A} |a| \geq p^n$, with equality only when $|a| = p$ for each $a \in A$. In this case, also, the lemma holds. So for a given n we will use an induction on $h(A)$.

If for each nonempty subset $B \subset A$, $B \neq A$, $|\langle B \rangle| > p^{|B|}$ replace one element of A by its pth power to get the set A'. The conditions of the lemma are satisfied for A' and we have $h(A') < h(A)$. Noting that $\langle A' \rangle = \langle A \rangle$, by induction on $h(A)$ we see that the lemma holds for A. Now assume, on the other hand, that there is a nonempty proper subset B of A such that $|\langle B \rangle| = p^{|B|}$. Clearly, B satisfies the conditions of the lemma and $|B| < |A|$. By the inductive assumption on the cardinality of A, $\langle B \rangle$ has a desired factorization

$$\langle B \rangle = \prod_{b \in B} \{e, b^{s(b)}, b^{2s(b)}, \ldots, b^{(p-1)s(b)}\},$$

where $s(b)$ is a power of p. Consider the factor group $G' = \langle A \rangle / \langle B \rangle$ and the set of cosets $A' = \{a\langle B \rangle : a \in A \setminus B\}$. A straightforward computation shows that $|\langle A' \rangle| = p^{|A'|}$ and that $|\langle B' \rangle| \geq p^{|B'|}$ for all $B' \subset A'$. By induction on the cardinality of A, we obtain a factorization of $\langle A' \rangle$ of the type in the statement of the lemma. Using the factorizations of $\langle B \rangle$ and $G' = \langle A \rangle / \langle B \rangle$, we obtain the desired factorization of $\langle A \rangle$. $\qquad\square$

Exercise 6. Verify Lemma 2 when $|A| = 1$.

Exercise 7. Verify Lemma 2 when $|a| = p$ for each $a \in A$.

Exercise 8. Fill in the details in the preceding proof.

With the aid of Lemmas 1 and 2 we now prove Hajós's theorem for p-groups.

Lemma 3. *Let G be a finite abelian p-group, and $G = A_1 A_2 \cdots A_n$ be a factorization, where $A_i = \{e, a_i, a_i^2, \ldots, a_i^{p-1}\}$. Then at least one of the A_i is a subgroup of G.*

Proof. Let

$$G = A_1 \cdots A_n \qquad (3)$$

be a factorization of the finite abelian p-group, where

$$A_i = \{e, a_i, a_i^2, \ldots, a_i^{p-1}\}.$$

Lemma 2 is applicable to $A = \{a_1, \ldots, a_n\}$. So for each i, $1 \leq i \leq n$ there is a power of p, say $s(i)$, and a subset

$$A_i' = \{e, a_i^{s(i)}, a_i^{2s(i)}, \ldots, a_i^{(p-1)s(i)}\}$$

such that $G = A'_1 \cdots A'_n$ is a factorization of G and at least one of the factors A'_i is a subgroup of G.

If $s(i) = 1$ for each i, $1 \leq i \leq n$, then $A_i = A'_i$ and we are done. So we assume that $s(i) \neq 1$ for some i. We index the elements of A such that $s(1) \neq 1, \ldots, s(m) \neq 1, s(m+1) = \cdots = s(n) = 1$ and $m \geq 1$. Since $s(i) = 1$ for each i, $m + 1 \leq i \leq n$, we have that

$$G = A'_1 \cdots A'_m A_{m+1} \cdots A_n$$

is a factorization of G. Hence the element $a_1 \cdots a_m$ can be represented in the form

$$a_1 \cdots a_m = a_1^{s(1)t(1)} \cdots a_m^{s(m)t(m)} a_{m+1}^{t(m+1)} \cdots a_n^{t(n)},$$

where $0 \leq t(i) \leq p - 1$. So

$$e = a_1^{s(1)t(1)-1} \cdots a_m^{s(m)t(m)-1} a_{m+1}^{t(m+1)} \cdots a_n^{t(n)}. \tag{4}$$

Now $s(i)t(i) - 1$ is relatively prime to p for each i, $1 \leq i \leq m$. So by Lemma 1, A_i can be replaced by

$$A_i^* = \left\{ e, a_i^{s(i)t(i)-1}, a_i^{2(s(i)t(i)-1)}, \ldots, a_i^{(p-1)(s(i)t(i)-1)} \right\}$$

in the factorization (3) to get the factorization

$$G = A_1^* \cdots A_m^* A_{m+1} \cdots A_n.$$

Equation (4) violates this factorization unless $m = 0$. This completes the proof.   □

2. Some examples and the general replacement principle

To make the proof of Rédei's theorem more accessible we pause to provide examples to introduce some of the ideas that appear in it.

Exercise 9.   Show that if G is a 2-group, then Rédei's theorem is a consequence of Hajós's theorem, in the special case described in Lemma 3.

We introduce the notion of *type* of a finite abelian group. If G is a direct product of cyclic groups of order $m_1, \ldots, m_t$, we say that G is of type $(m_1, \ldots, m_t)$. A group may have different types. For instance the cyclic group of order 6 is of types (6), $(2, 3)$, and $(3, 2)$. (If we insist that each m_i is a prime power, the type of a group would be unique up to order.)

Consider groups G of the types (9) or $(3, 3)$. Take the first case, when G is the cyclic group of order 9 with generator g. Suppose that $G = A_1 A_2$ is a factorization of G, where $A_1 = \{e, g^a, g^b\}$ and $A_2 = \{e, g^c, g^d\}$. Here we may assume that $1 \leq a, b, c, d \leq 8$. If $\{a, b\}$ or $\{c, d\}$ is the set $\{3, 6\}$, then A_1 or A_2 is a subgroup of G.

Exercise 10.   How many choices do we have for a, b, c, d to verify Rédei's theorem for the cycle group of order 9 by brute force?

The reader who attempts to verify Rédei's theorem in the style of Exercise 10 will appreciate the need for some devices to simplify the bookkeeping. The basic law of exponents, $x^a x^b = x^{a+b}$, suggests that we introduce polynomials. (This idea goes back to Euler who used formal power series $\sum_{n=0}^{\infty} f(n) x^n$ to record a function f defined on the nonnegative integers.)

The factorization $G = A_1 A_2$ can be expressed in terms of the product

$$(1 + x^a + x^b)(1 + x^c + x^d). \tag{5}$$

When you expand (5), you will get a polynomial $P(x)$ with nine terms, each of which has coefficient 1. Furthermore, the nine exponents are distinct modulo 9. If an exponent is larger than 8, we can reduce it as follows. Say that x^{14} appears in $P(x)$. Write 14 as $1 \cdot 9 + 5$. Then $x^{14} = x^9 x^5$, hence $x^{14} - x^5 = (x^9 - 1)x^5$ which shows that x^{14} is congruent to x^5 modulo $x^9 - 1$. This type of reduction implies that there is a polynomial $Q(x)$ with integer coefficients such that

$$(1 + x^a + x^b)(1 + x^c + x^d) = 1 + x + x^2 + \cdots + x^8 + Q(x)(x^9 - 1).$$

Rather than deal with this identity let us replace x by $\rho = e^{2\pi i/9}$, a primitive 9th root of 1. We then obtain an equation in the field of complex numbers,

$$(1 + \rho^a + \rho^b)(1 + \rho^c + \rho^d) = 0. \qquad (6)$$

At least one of the factors in (6) must be 0. Assume, say, that $1 + \rho^a + \rho^b = 0$. Drawing three ninth roots of unity 1, ρ^a, and ρ^b in the complex plane we can convince ourselves that $\{a, b\} = \{3, 6\}$ is the only case when their sum is 0.

Looking back on this argument, we see that we could have omitted any reference to polynomials, and gone directly from the factorization $G = A_1 A_2$ to the equation

$$(1 + \rho^a + \rho^b)(1 + \rho^c + \rho^d) = 1 + \rho + \rho^2 + \cdots + \rho^8. \qquad (7)$$

In fact, (7) holds for any of the nine 9th roots of 1. For instance, replacing ρ by 1, we obtain the equation $3 \cdot 3 = 9$, which comes as no surprise since the product of the orders of the factors equals the order of the group.

Even though the original problem concerns groups, which have only one composition, the solution exploits a ring of polynomials or the complex field, which are richer structures, possessing both a multiplication and an addition. The proof of Rédei's theorem uses both structures as well as characters of a finite abelian group.

A function χ from the abelian group G to the complex numbers of modulus 1 is called a *character* of G if $\chi(gh) = \chi(g)\chi(h)$ for each $g, h \in G$. The character for which $\chi(g) = 1$ for each $g \in G$ is called the *principal* character of G. For any subset A of G define $\chi(A)$ to be the sum of the numbers $\chi(a)$ for all $a \in A$. In particular, if χ is the principal character of G, $\chi(A) = |A|$. (Appendix B develops the basic properties of characters.)

Exercise 11. Let $G = A_1 A_2$ be a factorization and χ be a character of G. Prove that $\chi(G) = \chi(A_1)\chi(A_2)$.

We repeat the previous argument using the terminology of characters.

Let ρ be a primitive 9th root of unity. Define the character χ of G by setting $\chi(g) = \rho$ and therefore $\chi(g^i) = \rho^i, 0 \leq i \leq 8$.

Applying this character to the factorization $G = A_1 A_2$, we obtain:

$$\chi(G) = \chi(e) + \chi(g) + \chi(g^2) + \cdots + \chi(g^8)$$
$$= 1 + \rho + \rho^2 + \cdots + \rho^8 = 0$$
$$\chi(A_1) = \chi(e) + \chi(g^a) + \chi(g^b) = 1 + \rho^a + \rho^b,$$
$$\chi(A_2) = \chi(e) + \chi(g^c) + \chi(g^d) = 1 + \rho^c + \rho^d,$$

and

$$\chi(G) = \chi(A_1)\chi(A_2).$$

From $0 = \chi(G) = \chi(A_1)\chi(A_2)$ it follows that $\chi(A_1) = 0$ or $\chi(A_2) = 0$. For the sake of definiteness suppose that $\chi(A_1) = 0$. Hence $\{a, b\} = \{3, 6\}$.

Exercise 12. Show that if G is the cyclic group of order 3^n and $G = A_1 \cdots A_n$ is a normed factorization of G, where $|A_1| = \cdots = |A_n| = 3$, then one of the factors $A_1, \ldots, A_n$ is a subgroup of G.

In the next few examples we will develop this tool further.

First we use characters to establish Rédei's theorem for any cyclic p-group.

Let p be a prime and let G be the cyclic group of order p^n with generator g. Suppose that $G = A_1 \cdots A_n$ is a normed factorization of G, where $|A_1| = \cdots = |A_n| = p$. Consider a primitive (p^n)th root of unity ρ and define the character χ of G by setting $\chi(g) = \rho$. Now as before there is a factor, say

$$A_1 = \{e, g^{a_1}, g^{a_2}, \ldots, g^{a_{p-1}}\}$$

for which

$$0 = \chi(A_1) = \chi(e) + \chi(g^{a_1}) + \chi(g^{a_2}) + \cdots + \chi(g^{a_{p-1}}) =$$
$$= 1 + \rho^{a_1} + \rho^{a_2} + \cdots + \rho^{a_{p-1}} = 0.$$

We may assume that $1 \leq a_i \leq p^n - 1$ for each i, $1 \leq i \leq p - 1$ We want to show that A_1 is a subgroup of G, that is, the exponents $0, a_1, a_2, \ldots, a_{p-1}$ form a permutation of the numbers $0, p^{n-1}$,

$2p^{n-1}, \ldots, (p-1)p^{n-1}$. To do this consider the polynomial $A_1(x)$ associated with A_1, defined by

$$A_1(x) = 1 + x^{a_1} + x^{a_2} + \cdots + x^{a_{p-1}}.$$

Next we need the fact that the (p^n)th cyclotomic polynomial

$$F(x) = 1 + x^{p^{n-1}} + x^{2p^{n-1}} + \cdots + x^{(p-1)p^{n-1}}$$

is irreducible over the field of rational numbers [2, p.46]. (Appendix D discusses the cyclotomic polynomials.) Note that $F(x)$ and $A_1(x)$ have a common root, namely ρ. Thus $F(x)$ divides $A_1(x)$ in the ring of polynomials with rational coefficients, and since $F(x)$ is monic, in the ring of polynomials with integer coefficients. The next exercise completes the argument.

Exercise 13. Show that from the fact that $F(x)$ divides $A_1(x)$ it follows that $0, a_1, a_2, \ldots, a_{p-1}$ is a permutation of $0, p^{n-1}, 2p^{n-1}, \ldots, (p-1)p^{n-1}$. (Hint: Let $A_1(x) = F(x)Q(x)$ and first show that the degree of $Q(x)$ is less than p^{n-1}. Then show that $Q(x)$ must be 1.)

We summarize this argument in Lemma 4, which will play a role in the proof of Rédei's theorem.

Lemma 4. *Let G be the cyclic group of order p^n where p is a prime and let $G = A_1 A_2 \cdots A_n$ be a normed factorization. Then at least one set A_i is a subgroup of G.*

The most important result about vanishing sums of roots of unity we use in connection with factoring p-groups generalizes a previous observation.

Lemma 5. *Let p be a prime and ρ a (p^n)th root of unity. If $\rho^{a_1} + \cdots + \rho^{a_s} = 0$, where $a_1 = 0$ and $1 \le s \le p$, then $s = p$ and $\rho^{a_1}, \ldots, \rho^{a_p}$ is a permutation of $1, \theta, \ldots, \theta^{p-1}$, where $\theta = e^{2\pi i/p}$.*

Exercise 14. Prove the lemma, using the technique in Exercise 13.

We have seen that characters are of use in the study of factorization because if

$$G = A_1 A_2 \cdots A_n,$$

then

$$\chi(G) = \chi(A_1)\chi(A_2) \cdots \chi(A_n).$$

But characters convey much more information, and we will use the following fact: If A and B are subsets of the finite abelian group G and if $\chi(A) = \chi(B)$ for each character χ of G then $A = B$. (That $\chi(A) = \chi(B)$ for the principal character tells us that $|A| = |B|$.)

An example will illustrate this observation. Let G be the cyclic group of order four. Denote a generator of G by g and a primitive fourth root of unity by ρ, which could be chosen to be i. The four characters of G, $\chi_1, \chi_2, \chi_3, \chi_4$ are given in Table 1.

	e	g	g^2	g^3
χ_1	1	1	1	1
χ_2	1	ρ	ρ^2	ρ^3
χ_3	1	ρ^2	1	ρ^2
χ_4	1	ρ^3	ρ^2	ρ

TABLE 1

Exercise 15.   Show that the columns of Table 1 are linearly independent. (Appendix B shows that the columns of such a character table are always linearly independent.)

Let us check the claim that if $\chi(A) = \chi(B)$ for each character χ of G, then $A = B$. From $\chi_1(A) = \chi_1(B)$ it follows that $|A| = |B|$. If for example $|A| = |B| = 2$ and $A = \{a_1, a_2\}$, $B = \{b_1, b_2\}$ then the fact that $\chi(A) = \chi(B)$ for each character χ of G is equivalent to the following four equations

$$\chi_1(a_1) + \chi_1(a_2) = \chi_1(b_1) + \chi_1(b_2)$$

$$\chi_2(a_1) + \chi_2(a_2) = \chi_2(b_1) + \overset{v}{\chi}_2(b_2)$$

$$\chi_3(a_1) + \chi_3(a_2) = \chi_3(b_1) + \chi_3(b_2)$$

$$\chi_4(a_1) + \chi_4(a_2) = \chi_4(b_1) + \chi_4(b_2).$$

These equations imply that a linear combination of two columns of Table 1 is equal to a linear combination of two columns of Table 1. Since the columns of Table 1 are linearly independent over the field of complex numbers we conclude that $A = B$.

So far we have extended the domain of a character from elements of a group G to subsets of G. Now it is necessary to extend the domain further.

Let $G = \{g_1, g_2, \ldots, g_n\}$ be a finite abelian group and consider the set of expressions of the form

$$z_1 g_1 + z_2 g_2 + \cdots + z_n g_n, \tag{8}$$

where $z_i \in \mathbb{Z}$. (Such an expression is shorthand for the function that assigns to g_i the integer z_i. See Appendix C.) We add two such expressions coordinate by coordinate:

$$(z_1 g_1 + z_2 g_2 + \cdots + z_n g_n) + (t_1 g_1 + t_2 g_2 + \cdots + t_n g_n)$$

$$= (z_1 + t_1)g_1 + (z_2 + t_2)g_2 + \cdots + (z_n + t_n)g_n.$$

The product of two such expressions is defined using the multiplication in G:

$$\left(\sum_{i=1}^{n} z_i g_i\right)\left(\sum_{j=1}^{n} t_j g_j\right) = \sum_{k=1}^{n} \sum_{g_i g_j = g_k} z_i t_j g_k.$$

This is similar to multiplication of polynomials.

The set of expressions (8), together with the addition and multiplication just introduced, is called the *group ring* of G; and is denoted $Z(G)$.

A subset A in G can be identified with the element in the group ring whose coefficients are 0 or 1: 1 at elements in A and 0 at elements in $G \setminus A$.

Exercise 16. Let G be the cyclic group of order 5 with generator u and identity element e. In $Z(G)$ compute
(a) $(e - u)(e + u + u^2 + u^3 + u^4)$,
(b) $(e + 2u + u^2)(3u^2 - u^4)$.

Now let A and B be subsets of a finite abelian group G. Considering A and B as elements of the group ring $Z(G)$, we may compute their product AB, obtaining an element $\sum_{g \in G} n(g)g$. The coefficient $n(g)$ records the number of ordered pairs (a, b) such that $ab = g$. That $G = AB$ is a factorization is equivalent to the assertion that $n(g) = 1$ for every $g \in G$. That the symbol AB has two meanings, namely a product in the group ring and a factoring of G will cause no difficulty. The context will make it clear which is intended.

Exercise 17. Let $z = \sum z_i g_i$ and $t = \sum t_i g_i$ be two elements in $Z(G)$ and χ a character on G. Prove that
(a) $\chi(zt) = \chi(z)\chi(t)$,
(b) $\chi(z + t) = \chi(z) + \chi(t)$.

Exercise 18.
(a) Let $z \in Z(G)$ have the property that $\chi(z) = 0$ for all characters χ on G. Prove that $z = 0$, the zero element of the ring $Z(G)$.
(b) Prove that if z and t are in $Z(G)$ and $\chi(z) = \chi(t)$ for all characters χ on G, then $z = t$.

As Exercises 17 and 18 show, the characters of a group G convey a great deal of information. In fact, as we will now show, they even can tell us whether two subsets of G, A and B, provide a factorization of G.

First of all, if χ is a nonprincipal character, $\chi(G) = 0$, as is shown in Appendix B. If χ is the principal character of G, $\chi(G) = |G|$. Now assume that $G = AB$ is a factorization. Then if χ is a nonprincipal character of G, at least one of the numbers $\chi(A)$ and $\chi(B)$ is 0. (If χ is the principal character of G, the equation $\chi(G) = \chi(A)\chi(B)$ says only that $|G| = |A||B|$.)

Now consider *any* subsets A and B of G. Then $G = AB$ is a factorization if and only if $G = AB$ in $Z(G)$. But $G = AB$ in $Z(G)$ if and only if $\chi(G) = \chi(AB)$ for all characters χ of G. This is equivalent to the assertion that $\chi(G) = \chi(A)\chi(B)$ for all characters of G. And this reduces to the statement "$G = AB$ *is a factorization if and only if* $|G| = |A||B|$ *and for every nonprincipal character* χ *of* G *at least one of* $\chi(A)$ *and* $\chi(B)$ *is* 0." This observation is the basis of the following replacement principle, which will be used several times in the proof of Rédei's theorem.

General Replacement Principle. *Let A and A' be subsets of G such that $|A| = |A'|$ and if $\chi(A) = 0$ it follows that $\chi(A') = 0$. Then A can be replaced by A' in all factorizations of G; that is, if $G = AB$ is a factorization, so is $G = A'B$.*

Exercise 19. Prove the General Replacement Principle.

Exercise 20. Show that the first replacement principle (Lemma 1) is a consequence of the General Replacement Principle.

Though we used the group ring to develop the General Replacement Principle, the group ring does not appear in its statement. We could have avoided mentioning the group ring, but only by using some clumsy bookkeeping. We would have to define the "product" AB of two subsets A and B as a set in which each element is assigned a weight, namely the number of times it is represented in the form ab, $a \in A$, $b \in B$. Then we would show that $\chi(AB) = \chi(A)\chi(B)$. It is far more natural to introduce the ring $Z(G)$ and observe that each character provides a homomorphism from it to the ring of complex numbers. (Moreover, when studying multiple tilings of R^n, the group ring is essential, for the elements in it with negative coeffients and zero divisors play a role.)

As our first application of the General Replacement Principle we have the following replacement principle of use in p-groups.

Lemma 6. (Second Replacement Principle) *Let A be a subset of the finite abelian p-group G such that* $|A| = p$ *and* $e \in A$. *Let k be relatively prime to p. Then A can be replaced by*

$$A' = \{e, a^k, a^{2k}, \ldots, a^{(p-1)k}\}$$

in each factorization of G for each $a \in A \setminus \{e\}$.

Exercise 21. Using Lemma 5 and the General Replacement Principle, prove Lemma 6.

Problem 1. The proof of Lemma 6 involves the use of characters, though characters are not mentioned in it. Obtain a direct proof, one that does not use characters or complex numbers.

Lemma 7. (Third replacement principle.) *Let A have p elements, where p is prime. Assume that each element in A has order 1 or p. Then A can be replaced by a cyclic subset in any factorization.*

Exercise 22. Prove Lemma 7.

Let us illustrate the use of replacement techniques in the factorings of a group of type $(3, 3)$, the non-cyclic group G of order 9.

Suppose that $G = A_1 A_2$ is a normed factorization of G, where $|A_1| = |A_2| = 3$. We will use characters to show that A_1 or A_2 is a subgroup of G. There are simpler ways to prove this but our purpose is to shed light on the methods we will use later.

Let ρ be a primitive third root of unity and h and k a basis of G. Each character χ of G is determined by $\chi(h)$ and $\chi(k)$. Both $\chi(h)$ and $\chi(k)$ can be assigned only three possible values, 1, ρ, or ρ^2. For the sake of concreteness we display the nine characters of G in Table 2.

For each nonprincipal character χ of G we have $\chi(A_1) = 0$ or $\chi(A_2) = 0$. On the other hand, if A_1 and A_2 are subsets of G such that $|A_1| = |A_2| = 3$ and for each nonprincipal character χ of G, $\chi(A_1) = 0$ or $\chi(A_2) = 0$, then we have the factorization $G = A_1 A_2$.

	e	h	h^2	k	hk	h^2k	k^2	hk^2	h^2k^2
χ_1	1	1	1	1	1	1	1	1	1
χ_2	1	ρ	ρ^2	1	ρ	ρ^2	1	ρ	ρ^2
χ_3	1	ρ^2	ρ	1	ρ^2	ρ	1	ρ^2	ρ
χ_4	1	1	1	ρ	ρ	ρ	ρ^2	ρ^2	ρ^2
χ_5	1	ρ	ρ^2	ρ	ρ^2	1	ρ^2	1	ρ
χ_6	1	ρ^2	ρ	ρ	1	ρ^2	ρ^2	ρ	1
χ_7	1	1	1	ρ^2	ρ^2	ρ^2	ρ	ρ	ρ
χ_8	1	ρ	ρ^2	ρ^2	1	ρ	ρ	ρ^2	1
χ_9	1	ρ^2	ρ	ρ^2	ρ	1	ρ	1	ρ^2

TABLE 2

Now let

$$A_1 = \{e, a_{11}, a_{12}\} \quad \text{and} \quad A_2 = \{e, a_{21}, a_{22}\}.$$

If χ is a character of G for which $\chi(A_1) = 0$, then $\chi(e)$, $\chi(a_{11})$, $\chi(a_{12})$ must be a permutation of 1, ρ, ρ^2. It follows that if $\chi(A_1) = 0$ then $\chi(H) = 0$, where $H = \{e, a_{11}, a_{11}^2\}$. Thus $G = HA_2$ is also a factorization of G. H is a subgroup of G and so A_2 is a complete set of representatives of the cosets of H ("modulo H" for short). Similarly, $G = A_1K$ is also a factorization of G, where $K = \{e, a_{21}, a_{21}^2\}$. K is a subgroup of G and so A_1 is a complete set of representatives modulo K. Note that $G = HK$ is also a factorization. This implies that a_{11} and a_{21} form a basis of G. For simplicity we denote $a_{11} = h$ and $a_{21} = k$. Thus

$$A_1 = \{e, h, h^2x\} \quad \text{and} \quad A_2 = \{e, k, k^2y\},$$

where $x \in K$, $y \in H$, $H = \{e, h, h^2\}$ and $K = \{e, k, k^2\}$. If $x = e$ or $y = e$, then A_1 or A_2 is a subgroup of G.

Exercise 23. Inspecting all the nine possible choices for x and y, verify that A_1 or A_2 is a subgroup of G.

Characters of G simplify the bookkeeping again. Suppose that $x \neq e$ and $y \neq e$. Pick a character χ of G for which $\chi(A_1) = 0$. We know that for this character $\chi(H) = 0$. Hence $\chi(h^2 x) = \chi(h^2)$, that is, $\chi(x) = 1$. Since x generates K, $\chi(k) = 1$. So there are at most three characters χ such that $\chi(A_1) = 0$. Similarly, there are at most three characters χ such that $\chi(A_2) = 0$. Thus there are at most $3 + 3 = 6$ characters χ of G for which $\chi(A_1 A_2) = 0$. However, this is not the case since there are 8 characters χ of G for which $\chi(A_1 A_2) = 0$.

Let us look at one more example before beginning the proof of Rédei's theorem. Let G be of type $(5, 5)$. Suppose that $G = A_1 A_2$ is a normed factorization of G. As before, A_1 can be replaced by the subgroup $H = \langle h \rangle$. Hence A_2 is a complete set of representatives modulo H. Similarly, A_2 can be replaced by the subgroup $K = \langle k \rangle$ and so A_1 is a complete set of representatives modulo K. Since $G = HK$ is also a factorization, h and k form a basis of G. Thus

$$A_1 = \{e, h, h^2 x, h^3 y, h^4 z\} \quad \text{and} \quad A_2 = \{e, k, k^2 u, k^3 v, k^4 w\},$$

where $x, y, z \in K$ and $u, v, w \in H$.

Exercise 24. How many choices do we have for the elements $x, y,$ z and u, v, w in order to verify Rédei's theorem by brute force? (Do not carry out the verification.)

Exercise 25. Let p be a prime and G be the group of type (p, p). Suppose that $G = A_1 A_2$ is a normed factorization, where $|A_1| = |A_2| = p$. Show that there is a basis, k and h of G, such that

$$A_1 = \{e, h, h^2 k^{a_2}, \ldots, h^{p-1} k^{a_{p-1}}\},$$
$$A_2 = \{e, k, k^2 h^{b_2}, \ldots, k^{p-1} h^{b_{p-1}}\}.$$

Rather than complete the case $(5, 5)$ we turn our attention to the general case (p, p), p prime, which is far from trivial.

3. Rédei's theorem for p-groups

With the tools we have developed so far we can prove Rédei's theorem for p-groups. The proof proceeds in two stages. First we treat the case when G is of type (p, p); then we prove the theorem for any finite abelian p-group. Recall that we have already proved it for groups of the type (p^n), that is, for cyclic p-groups.

The next exercise establishes some facts about polynomials whose coefficients lie in $GF(p)$, the field with p elements. These results will be used in the proof of Lemma 8.

Exercise 26. Let $E(x) = \prod_{i=1}^{n}(x - a_i)$, where $a_i \in GF(p)$. Let $c_1, c_2, \ldots, c_m$, $m \leq n$, be the distinct values occurring among the a_i's. Let $F(x) = \prod_{i=1}^{m}(x - c_i)$. Show that
(a) $F(x)$ divides $E(x)$ and $x^p - x$.
(b) $E(x)/F(x)$ divides $E'(x)$, the derivative of $E(x)$.

Lemma 8. *Rédei's theorem is true for groups of type (p, p), where p is a prime.*

Proof. Let G be of type (p, p) and suppose that $G = A_1 A_2$ is a normed factorization of G, where $|A_1| = |A_2| = p$. By Exercise 25 there is a basis k and h of G such that

$$A_1 = \{e, h, h^2 k^{a_2}, \ldots, h^{p-1} k^{a_{p-1}}\},$$
$$A_2 = \{e, k, k^2 h^{b_2}, \ldots, k^{p-1} h^{b_{p-1}}\}.$$

If $p = 2$, then both A_1 and A_2 are subgroups of G. From now on we suppose that p is odd.

We wish to show that all the a_i's are 0 modulo p or that all the b_j's are 0 modulo p. For instance, in the first case, we want to show that the polynomial $(x - a_2)(x - a_3) \cdots (x - a_{p-1})$, viewed as a polynomial over $GF(p)$, is just the polynomial x^{p-2}. Introducing $a_0 = a_1 = 0$, we want to show that $\prod_{i=0}^{p-1}(x - a_i)$ is simply x^p.

To begin the argument let ρ be a primitive pth root of unity. For each integer y, $0 \leq y \leq p - 1$ define the (nonprincipal) character χ_y on G by $\chi_y(h) = \rho^y$ and $\chi_y(k) = \rho^{-1-y}$. Recalling that p is odd, we see that $\chi_y(A_1) = 0$ for at least $(p + 1)/2$ values of y or $\chi_y(A_2) = 0$ for at least $(p + 1)/2$ values of y. We assume without loss of generality that the first case occurs. From this we will deduce that $a_2 = a_3 = \cdots = a_{p-1} = 0$.

We have then

$$0 = \chi_y(A_1) = \sum_{i=0}^{p-1} \rho^{iy - a_i(1+y)}$$

for at least $(p+1)/2$ values of y. Let Y be the set of these values of y. Since $\chi_y(A_1) = 0$, the exponents $iy - a_i(1+y)$ form a complete set of representatives modulo p for each $y \in Y$.

Let $c_0, c_1, \ldots, c_m$ be the distinct values occurring among a_0, $a_1, \ldots, a_{p-1}$, taken modulo p. We define three polynomials with coefficients in $GF(p)$:

$$D(x,y) = \prod_{i=0}^{p-1} (x + iy - a_i(1+y)) = \sum_{i=0}^{p} d_i(y)x^i,$$

$$E(x) = D(x,0) = \prod_{i=0}^{p-1} (x - a_i) = \sum_{i=0}^{p} e_i x^i,$$

$$F(x) = \prod_{i=0}^{m} (x - c_i),$$

and we wish to prove that $E(x) = x^p$, or equivalently, that $E'(x) = 0$.

We first show that many of the coefficients of $E(x)$ are 0, by examining $D(x,y)$.

Let us write

$$D(x,y) = d_0(y) + d_1(y)x + \cdots + d_p(y)x^p.$$

Since $D(x,y) = \prod_{j=0}^{p-1}(x - (a_j(1+y) - jy))$, we have $d_p(y) = 1$ and each $d_{p-i}(y)$, $0 \le i \le p-1$, is an elementary symmetric function in the p expressions $a_j(1+y) - jy$. Let $S_i(z_j)$ denote the ith elementary function of the expressions $z_0, \ldots, z_{p-1}$. Then we have

$$d_{p-i} = S_i(jy - a_j(1+y)).$$

Note that the degree of $d_{p-i}(y)$ is at most i.

Now $D(x,y) = -x + x^p$ for each y in Y. Thus for $y \in Y$ we have $d_{p-i}(y) = 0$ for $1 \le i \le p - 2$. Therefore for those values of i the polynomial $d_{p-i}(y)$ has at least $(p+1)/2$ roots. Since the degree of $d_{p-i}(y)$ is at most i, we see that for $1 \le i \le (p-1)/2$, $d_{p-i}(y)$ is the zero polynomial. That means that $D(x,y)$ has the form

$$D(x,y) = d_0(y) + d_1(y)x + \cdots + d_{(p-1)/2}(y)x^{(p-1)/2} + x^p.$$

Thus

$$E(x) = e_0 + e_1 x + \cdots + e_{(p-1)/2}x^{(p-1)/2} + x^p,$$

where $e_i \in GF(p)$. Moreover, since $a_0 = a_1 = 0$ and are roots of $E(x)$, we may write

$$E(x) = e_2 x^2 + \cdots + e_{(p-1)/2}x^{(p-1)/2} + x^p.$$

All that remains is to show that $e_2 = e_3 = \cdots = e_{(p-1)/2} = 0$. To do that, we show that $E'(x) = 0$.

Introduce $G(x) = E(x) - (x^p - x)$, a polynomial of degree at most $(p-1)/2$. It is not the zero polynomial since it has a term of degree 1. Since E' has degree at most $(p-3)/2$, the product $G(x)E'(x)$ has degree at most $p-2$ or is the zero polynomial. We show that $E(x)$, which has degree p, divides $G(x)E'(x)$.

By Exercise 26, $E(x)/F(x)$ divides $E'(x)$. Hence there is a polynomial $Q(x)$ such that

$$E'(x) = \frac{Q(x)E(x)}{F(x)},$$

Thus

$$G(x)E'(x) = \frac{G(x)}{F(x)}Q(x)E(x).$$

Since $F(x)$ divides $x^p - x$ and $E(x)$, it divides their difference, $G(x)$. Hence $E(x)$ divides $G(x)F'(x)$. Since the degree of $G(x)E'(x)$ is smaller than the degree of $E(x)$, $G(x)E'(x)$ is the zero polynomial, hence $E'(x) = 0$. Therefore $E(x) = x^p$ and the proof is complete.

□

In order to prove Rédei's theorem for any finite abelian p-group it will be useful to note that if Rédei's theorem is true, then the following seemingly sharper Lemma 9 also holds. The proof of this lemma depends on the following exercise.

Exercise 27.   Let G be a group and N a normal subgroup of G. Let B be a subset of G such that $\{bN : b \in B\}$ is a subgroup of G/N. Then $\cup_{b \in B}\{bN\}$ is a subgroup of G.

Lemma 9.   *Let $G = A_1 \cdots A_n$ be a normalized factorization of the finite abelian group G by subsets of prime orders. Assume Rédei's theorem is true. Then there is a rearrangement of the factors, $A_1, \ldots, A_n$, say $B_1, \ldots, B_n$, such that all the subsets $B_1, B_1B_2, \ldots, B_1B_2 \cdots B_n$ are subgroups of G.*

Exercise 28.   Use Exercise 27 to prove Lemma 9 for the cases $n = 3$ and $n = 4$.

We are now ready to obtain the main result in this section.

Lemma 10.   Rédei's theorem holds for any finite abelian p-group.

Proof.   Let G be a finite abelian p-group of order p^n and $G = A_1 \cdots A_n$ be a normed factorization of G, where $|A_1| = \cdots = |A_n| = p$. We wisn to prove that at least one A_i is a subgroup of G. The case $n = 1$ is trivial. The case $n = 2$ is settled already, since Lemma 4 takes care of G of the type (p^2) and Lemma 8 the type (p, p). We argue by induction, starting with the cases $n = 1$ and $n = 2$, and consider $n \geq 3$.

By Lemma 6 every factor A_i can be replaced by a cyclic subset. If each factor contains an element of order at least p^2, then using them we can construct a factorization of G consisting of non-subgroup cyclic subsets. This contradicts Lemma 3. Thus there exists a factor, say A_1, whose nonidentity elements all have order p. By Lemma 7, A_1 can be replaced by a subgroup H, which is generated by a nonidentity element of A_1, to get the factorization $G = HA_2 \cdots A_n$. Considering the factor group G/H we have the factorization $G/H = (A_2 H)/H \cdots (A_n H)/H$. By the inductive assumption and Lemma 9 there is a permutation $B_1, \ldots, B_n$ of the factors $H, A_2, \ldots, A_n$ such that $B_1, B_1 B_2, \ldots, B_1 B_2 \cdots B_n$ are subgroups of G and $B_1 = H$. To be concrete we will suppose that the permutation $B_2, \ldots, B_n$ of the factors $A_2, \ldots, A_n$ is the identity since this is only a matter of indexing the factors A_i. Consider the subgroup $K = HA_2 \cdots A_{n-1}$. Since the identity element e is an element of each factor, H and each A_i is a subset of K. If $A_1 \subset K$, then $K = A_1 A_2 \cdots A_{n-1}$ is a factorization of K. (Note that both K and $A_1 A_2 \cdots A_{n-1}$ have p^{n-1} elements.) By the inductive assumption, at least one of $A_1, A_2, \ldots, A_{n-1}$ is a subgroup of K and so of G.

So suppose that A_1 is not a subset of K. Then replace A_1 in the factorization of G by a subgroup L generated by an element of A_1 that is not in K. Since $L \not\subset K$, we have $K \cap L = \{e\}$.

From the factorization $G = LA_2 \cdots A_n$ it follows, by applying the induction assumption to the group G/L, there is a permutation $C_1, \ldots, C_n$ of the factors $L, A_2, \ldots, A_n$ such that the subsets $C_1, C_1 C_2, \ldots, C_1 C_2 \cdots C_n$ are subgroups of G and $C_1 = L$. There is an index j such that $C_2 = A_j$ and therefore LA_j is a subgroup of G. If $j \neq n$, then $K \cap LA_j = A_j$, being the intersection of two subgroups, is a subgroup of G. Therefore we may assume that LA_n is a subgroup of G.

Consider the intersection $K \cap LA_n$. If it were just $\{e\}$, then G would contain the direct product of the two subgroups K and LA_n, a group whose order would be p^{n+1}, which is greater than $|G|$. Thus $|K \cap LA_n| = p^2$ or p.

If $|K \cap LA_n| = p^2$, then $LA_n \subset K$, hence $L \subset K$, which is a contradiction. Hence $|K \cap LA_n| = p$. Consider now two cases: LA_n cyclic and LA_n of type (p, p).

Say that LA_n is cyclic. Then $K \cap LA_n$ is the unique subgroup of order p in LA_n, namely L. Since $L \not\subset K$, this case is ruled out.

Since LA_n is not cyclic, the nonidentity elements of A_n have order p. Consequently A_n can be replaced by a subgroup M generated by a nonidentity element of A_n to get the factorizations $G =$

$A_1 \cdots A_{n-1}M$ and $G = HA_2 \cdots A_{n-1}M$. Note that $G = KM$ is also a factorization of G and so $K \cap M = \{e\}$.

From the factorization $G = A_1 \cdots A_{n-1}M$ it follows that there is a permutation $D_1, \ldots, D_n$ of the factors $A_1, \ldots, A_{n-1}, M$ such that $D_1 = M$ and the subsets

$$D_1, D_1 D_2, \ldots, D_1 D_2 \cdots D_n$$

are subgroups of G. There is an index j, $1 \le j \le n - 1$ such that $D_2 = A_j$. Hence MA_j is a subgroup of G.

If $j \ne 1$, then $A_j \subset K$. Recalling that $K \cap M = \{e\}$, we have $K \cap MA_j = A_j$ and so A_j, being the intersection of two subgroups, is a subgroup of G.

If $j = 1$, we have that $N = MA_1$ is a subgroup of G.

Consider two cases: $A_n \subset N$ and $A_n \not\subset N$. In the first case $A_1 A_n \subset N$. Therefore, since both N and $A_1 A_n$ contain p^2 elements, $N = A_1 A_n$. By Lemma 8 at least one of A_1 and A_n is a subgroup of N, hence of G.

The other case is $A_n \not\subset N$. Then A_n can be replaced by a subgroup T such that $T \cap N = \{e\}$. We then have $G = A_1 A_2 \cdots A_{n-1}T$. Applying the induction to the group G/T shows that there is a subgroup of G of the form TA_j, where $1 \le j \le n - 1$. Consider $K \cap TA_j = HA_2 \cdots A_{n-1} \cap TA_j$. As argued before, if this intersection were just $\{e\}$, then G would contain a subgroup of p^{n+1} elements. The intersection cannot be TA_j since K does not contain T. Thus the intersection is a subgroup of TA_j with p elements. If $j \ne 1$, this intersection is A_j, establishing the theorem. If $j = 1$, then both TA_1 and $N = MA_1$ are subgroups of G. Recalling that $T \cap N = \{e\}$, we conclude that $TA_1 \cap N = A_1$, hence that A_1 is a subgroup of G. This completes the proof.   □

4. Vanishing sums of roots of unity

To prove Rédei's theorem for non-p-groups we need another special case of the General Replacement Principle. It depends on another property of vanishing sum of roots of unity, which is given in Lemma 11. The proof of this lemma uses the next exercise.

Exercise 29. Show that if ρ is a primitive (rs)th root of unity, where r and s are relatively prime, then ρ is a product of a primitive rth and a primitive sth root of unity. (Hint: Since r and s are relatively prime there are integers a and b such that $1 = ar + bs$. Hence $\rho^1 = \rho^{ar+bs} = \rho^{ar}\rho^{bs}$.)

Lemma 11. *Let ρ be a primitive nth root of unity and let p be the least prime factor of n. If $\rho^{a_1} + \cdots + \rho^{a_s} = 0$, $a_1 = 0$ and $1 \le s \le p$, then $s = p$ and there is a primitive pth root of unity θ such that $\rho^{a_1}, \ldots, \rho^{a_p}$ is a permutation of $1, \theta, \ldots, \theta^{p-1}$.*

Proof. We proceed by induction on n. The case $n = p^e$ is settled in Lemma 5.

Suppose that $n = p^e r$, where r is relatively prime to p. Let $\rho = \sigma\tau$, where σ and τ are (p^e)th and rth primitive roots of unity respectively. Now

$$0 = \sum_{i=1}^{s} \rho^{a_i} = \sum_{i=1}^{s} \sigma^{a_i}\tau^{a_i}.$$

Divide p^e into a_i to obtain the remainder b_i such that $0 \le b_i \le p^e - 1$. Let $b_1', \ldots, b_t'$ be the different numbers among $b_1, \ldots, b_s$. Then

$$0 = \sum_{i=1}^{s} \rho^{a_i} = \sum_{i=1}^{s} \sigma^{b_i}\tau^{a_i} = \sum_{i=1}^{t} \alpha_i \sigma^{b_i'},$$

where α_i's are nonempty sums of rth roots of unity. Consider the polynomial

$$A(x) = \sum_{i=1}^{t} \alpha_i x^{b_i'}. \tag{9}$$

Suppose that one of the coefficients α_i equals 0. Divide it by one of its terms to obtain a sum $\tau^{c_1} + \tau^{c_2} + \cdots + \tau^{c_k}$, where τ is an rth root of unity and $c_1 = 0$. Note that $k \le s$. Since this sum is 0, by induction, k must be equal to the least prime factor of r, hence $k > p$. Thus $s \ge k > p$, contradicting the assumption that $s \le p$.

We may therefore assume that none of α_i's is zero and so $A(x)$ is not the zero polynomial. Hence $0 \le \deg A(x) \le p^e - 1$. The (p^e)th cyclotomic polynomial

$$F(x) = \sum_{i=0}^{p-1} x^{ip^{e-1}}$$

is irreducible over the rth cyclotomic field since r is relatively prime to p^e. (For a proof see Appendix D.) Note that σ is a common root of $A(x)$ and $F(x)$. Hence $A(x)$ is a multiple of $F(x)$, that is, $A(x) = B(x)F(x)$ for a suitable polynomial $B(x)$ with coefficients from the rth cyclotomic field.

Since the degree of $F(x)$ is $(p-1)p^{e-1}$, equation (9) tells us that $\deg B(x) \le p^{e-1} - 1$. Therefore the nonzero terms of $B(x)$ occur among the nonzero terms of $A(x)$. Since $B(x)$ has at least one nonzero term, $A(x)$ has at least p terms. Thus $p \le t \le s$. On the other hand, $p \ge s$. Hence $p = s$. This completes the proof. □

To prove Rédei's theorem for groups that are not p-groups we need the concepts of p-part and $\bar{p}$-part of an element of G for a prime p. Let H be the p-component of G and let K be the complementary direct factor to H in G. Each $g \in G$ is uniquely expressible in the form $g = hk$, $h \in H$, $k \in K$. The elements h and k are the p-part and the $\bar{p}$-part of g and they will be denoted by $g(p)$ and $\bar{g}(p)$ respectively. Thus $g = g(p)\bar{g}(p)$.

Lemma 12. (Fourth replacement principle) *Let A be a factor of G such that $|A| = p$ is the least prime factor of $|G|$ and $e \in A$. Then A can be replaced in any factorization of G by $A' = \{a(p)(\bar{a}(p))^{s(a)} : a \in A\}$ for any choice of integer exponents $s(a)$.*

Proof. Since A is a factor of G, there is a subset B of G such that $G = AB$ is a factorization and $e \in B$. We first show that there is a character of G that vanishes on A.

If there were no such character, then for every nonprincipal character χ, $\chi(B) = 0$. For the principal character χ of G, we have $\chi(B) = |B|$. Thus every character of G vanishes on the group ring element $|B|G - |G|B$, hence $|B|G = |G|B$. But this is impossible since the coefficients of e on the two sides of the equation are different.

Now let χ be a character of G for which

$$0 = \chi(A) = \sum_{a \in A} \chi(a).$$

There exists a minimal integer n such that each $\chi(a)$ is a power of a fixed primitive nth root of unity. Since n is a divisor of $|G|$ the least prime factor of n is at least p. Thus by Lemma 11 there is a primitive pth root of unity θ such that $\{\chi(a) : a \in A\} = \{\theta^i : 0 \le i \le p-1\}$. This implies that $|A'| = p$ and that from $\chi(A) = 0$ it follows that $\chi(A') = 0$. So, by the General Replacement Principle, A can be replaced by A' in each factorization of G. $\qquad\square$

Exercise 30. Justify the claim that $\chi(A') = 0$ in the preceding proof.

At one point near the end of the proof of Rédei's theorem we will need the following lemma.

Lemma 13. (Fifth replacement principle) *Let A be a subset of G such that $|A| = p$ is an odd prime factor of $|G|$ and $e \in A$. Suppose that the nonidentity elements of A have order p or $2p$. Then A can be replaced in any factorization of G by $A' = \{a(p)(\overline{a}(p))^{s(a)} : a \in A\}$ for any integers $s(a)$.*

Proof. Let $A = \{a_1, a_2, \ldots, a_p\}$, with $a_1 = e$. Write each a_i in the form $a_i = a_i(2)a_i(p)$. Here $(a_i(2))^2 = e$ and $(a_i(p))^p = e$. Note that $a_1(2) = a_1(p) = e$. As argued in the proof of Lemma 12, there is a character χ of G such that $\chi(A) = 0$. For such a character we have

$$\chi(a_1(2))\chi(a_1(p)) + \chi(a_2(2))\chi(a_2(p)) + \cdots + \chi(a_p(2))\chi(a_p(p)) = 0.$$

Now, $\chi(a_i(2)) = \varepsilon_i = \pm 1$ and $\chi(a_i(p)) = \rho^{n_i}$, where n_i may be chosen in the range from 0 to $p-1$ and ρ is a primitive pth root of

1. We have

$$\varepsilon_1 \rho^{n_1} + \varepsilon_2 \rho^{n_2} + \cdots + \varepsilon_p \rho^{n_p} = 0,$$

where $\varepsilon_1 = 1$ and $n_1 = 0$. Thus ρ is a root of

$$g(x) = \varepsilon_1 x^{n_1} + \varepsilon_2 x^{n_2} + \cdots + \varepsilon_p x^{n_p},$$

a polynomial of degree less than p, or the zero polynomial.

If it were the zero polynomial, the coefficient of x^{n_i}, which is a sum of 1's and (-1)'s would be 0. Hence there would be an even number of terms corresponding to each exponent. Since there are p terms, and p is odd, this would be a contradiction. Thus $g(x)$ is not the zero polynomial.

Since $g(\rho) = 0$ and $g(x)$ has constant term 1, $g(x) = F_p(x)$, the cyclotomic polynomial for the pth roots of 1. That implies that $\varepsilon_i = 1$ for all i and that $n_1, n_2, \ldots, n_p$ is a rearrangement of $0, 1, \ldots, p-1$. From this it follows that $|A'| = |A|$. Note that if $\chi(A) = 0$ then $\chi(A') = 0$. By the General Replacement Principle, A may be replaced by A' in any factorization. $\square$

5. Proof of Rédei's theorem

We come now to the proof of Rédei's theorem.

Let p be a prime divisor of the order of the finite abelian group G. We will say that an element $g \in G$ is a *p-element* if $\overline{g}(p) = e$. In other words, the p-elements are the elements in the p-component of G.

The proof also uses the result in the next exercise.

Exercise 31. Let $G = A_1 \cdots A_n$ be a normalized factorization of the finite abelian group G and suppose that each $|A_i|$ is prime. Let p be a prime divisor of $|G|$. Show that if all the factors of cardinality p contain only p-elements then these factors form a factorization of the p-component of G. (Hint: Consider cardinalities.)

Theorem 1. (Rédei's Theorem) *Let G be a finite abelian group and $G = A_1 A_2 \cdots A_n$ a normalized factorization such that each A_i con-*

tains a prime number of elements. Then at least one A_i is a subgroup of G.

Proof. By Lemma 10 the theorem holds for p-groups, so let G be a non-p-group and $G = A_1 \cdots A_n$ a normed factorization of G, where the factors are of prime cardinalities. We use the inductive assumption that the theorem holds for each proper subgroup H of G.

Let p be a prime divisor of $|G|$ and suppose that $A_1, \ldots, A_t$ are the factors in the factorization $G = A_1 \cdots A_n$ with cardinality p. The smallest value of

$$h_p(A_1, \ldots, A_n) = \prod_{i=1}^{t} \prod_{a \in A_i} |\bar{a}(p)|$$

is 1, which is attained only when each A_i, $1 \leq i \leq t$ contains only p-elements. In this case, by Exercise 31, $A_1 \cdots A_t$ is a factorization of the p-component of G. By Rédei's theorem for p-groups, at least one A_i is a subgroup of the p-component of G, hence of G. We now proceed by induction on $h_p(A_1, \ldots, A_n)$.

Let p be the least prime factor of $|G|$. We may assume that $h_p(A_1, \ldots, A_n) > 1$, that is, there is an i, $1 \leq i \leq t$ such that A_i contains not only p-elements. Let A_1 be such a factor and suppose $a \in A_1$ is a non-p-element. Thus there is a prime q such that $q \neq p$ and $a(q) \neq e$. Note that $q > p$.

Suppose that A_1 contains more than one non-p-element. By Lemma 12, A_1 can be replaced by a nonsubgroup A_1' such that $h_p(A_1', A_2, \ldots, A_n) < h_p(A_1, \ldots, A_n)$. (Let all but one of the exponents $s(a)$ in Lemma 12 be 0.) By the inductive assumption one of the factors $A_1', A_2, \ldots, A_n$ is a subgroup of G. Since A_1' is not a subgroup, at least one of $A_2, \ldots, A_n$ is a subgroup of G, and the theorem is established in this case.

Thus we may assume that a is the only non-p-element of A_1. Assume that $|a(p)| > p$. Then, by Lemma 12, A_1 can be replaced by A_1', the set which consists of the p-parts of the elements of A_1. (Let all the exponents $s(a)$ in Lemma 12 be 0.) Now, A_1' is not a subgroup of G and $h_p(A_1', A_2, \ldots, A_n) < h_p(A_1, \ldots, A_n)$. By the inductive assumption one of the factors $A_2, \ldots, A_n$ is a subgroup of G. We may suppose therefore that $|a(p)| = p$. Similarly, if $|\bar{a}(p)| > q$, then A_1 can be replaced by a nonsubgroup A_1' for which $h_p(A_1', A_2, \ldots, A_n) < h_p(A_1, \ldots, A_n)$. Once again by the inductive assumption one of the factors $A_2, \ldots, A_n$ is a subgroup of G. So, finally, we suppose that $|a| = pq$.

By Lemma 12, for each integer i, $1 \leq i \leq q - 1$, A_1 can be replaced by

$$A_{1,i} = \{a(p)(a(q))^i\} \cup (A_1 \setminus \{a\})$$

to get the factorization $G = A_{1,i}A_2 \cdots A_n$. We will use this fact soon. By Lemma 12 A_1 can be replaced by A_1', which consists of the p-parts of the elements of A_1. Now

$$h_p(A_1', A_2, \ldots, A_n) < h_p(A_1, \ldots, A_n).$$

By the inductive assumption one of the factors $A_1', A_2, \ldots, A_n$ is a subgroup of G. If this is not A_1', then we are done. Hence we assume that $H = A_1'$ is a subgroup of G.

From the factorization $G = HA_2 \cdots A_n$ it follows that there is a permutation $H = B_1, B_2, \ldots, B_n$ of the factors $H, A_2, \ldots, A_n$ such that $B_1, B_1B_2, \ldots, B_1B_2 \cdots B_n$ are subgroups of G. We may assume that the permutation is just the identity as this is only a matter of reindexing the factors.

Consider the subgroup $K = HA_2 \cdots A_{n-1}$. If $A_1 \subset K$, then $K = A_1A_2 \cdots A_{n-1}$ is a factorization of K. By the inductive assumption about the subgroups of G, one of the factors $A_1, \ldots, A_{n-1}$ is a subgroup of K, and so of G.

Thus we assume that $A_1 \not\subset K$, that is, $a(q) \notin K$. From this it follows that $|G : K| = q$. So from the factorization $G = KA_n$ we have $|A_n| = q$. Again, the factorization $G = KA_n$ implies that A_n is a complete set of representatives modulo K. Therefore for each i, $1 \leq i \leq q-1$ there is an $a_i \in A_n$ such that the coset a_iA_n contains the element $(a(p))^{-1}(a(q))^{-i}$. This means that $a(p)(a(q))^i a_i \in K$. Let

$$C_i = \{a(p)(a(q))^i a_i\} \cup (A_1 \setminus \{a\}) = \{a(p)(a(q))^i a_i\} \cup (H \setminus \{a(q)\}).$$

Note that $K = C_i A_2 \cdots A_{n-1}$ is a factorization of K. Indeed, products coming from $C_i A_2 \cdots A_{n-1}$ occur among the products coming from $A_{1,i} A_2 \cdots A_{n-1} A_n$ and these latter are distinct since $A_{1,i} A_2 \cdots A_{n-1} A_n$ is a factorization of G.

By the inductive assumption about the subgroups of G we have that one of the factors $C_i, A_2, \ldots, A_{n-1}$ is a subgroup of K. If this is not C_i we are done. Hence we assume that C_i is a subgroup of K.

We distinguish two cases depending on whether $p = 2$ or $p \geq 3$.

If $p \geq 3$, then the subgroups C_i and H have a nonidentity element in common and, both have prime order, $C_i = H$. Thus $a(p)(a(q))^i a_i = a(p)$, that is, $a_i = (a(q))^{-i}$. This means that $A_n = \langle a(q) \rangle$ and therefore A_n is a subgroup of G.

Finally we suppose that $p = 2$. Now $(a(p)(a(q))^i a_i)^2 = e$, hence $a_i(q) = (a(q))^{-i}$. Thus $(a(q))^{-i}$ are the q-parts of the non-identity elements of A_n. If the 2-part of each element of A_n is the identity element then A_n is a subgroup and we are done. Thus the order of each nonidentity element of A_n is either q or $2q$ and A_n contains an element whose order is $2q$.

Making use of the replacement principle in Lemma 13, we can repeat the whole argument, starting with A_n instead of A_1. Since $|A_n| = q \geq 3$ the procedure will not go on forever. This completes the proof. $\qquad\qquad\qquad\qquad\qquad\qquad\qquad\qquad\qquad\qquad\qquad\qquad$ $\square$

Problem 2.   Obtain a shorter or simpler proof of Rédei's theorem.

References

1. K. Corrádi and S. Szabó, A new proof of Rédei's theorem, *Pacific J. Math.* **140** (1989), 53–61.

2. D. S. Dummit and R. M. Foote, *Abstract Algebra*, Prentice Hall, Englewood Cliffs, 1991 (p. 466, irreducibility of the cyclotomic polynomials).

3. L. Rédei, Die neue Theorie der endlichen Abelschen Gruppen und Verallgemeinerung des Hauptsatzes von Hajós, *Acta Math. Acad. Sci. Hung.* **16** (1965), 329–373.

Epilog

While the central theme of the seven chapters has been the interplay of geometry with algebra, another theme, purely algebraic, also ran through the book. Time and time again we have used mappings from one structure to another to simplify a problem.

In Chapters 1 and 2, where we faced an infinite lattice L imbedded in a rational lattice L', we formed the quotient group L'/L. The natural homomorphism from L' to L'/L enabled us to work in L'/L, which is a finite group, instead of an infinite group.

At the end of Chapter 2 the solution of a brick-tiling problem depended on the homomorphism from the ring of polynomials $Z[x, y]$ to the ring of complex numbers obtained by replacing x and y by a root of unity.

Two homomorphisms were the key to Chapters 3 and 4 on the semicross and cross. One homomorphism, defined on Z^n, was onto a finite group G. It enabled us to lift a splitting of G to a tiling of Z^n. One way to construct such a splitting when G is a cyclic group is to find a "logarithm" $\phi : \{1, 2, \ldots, k\} \to C(k)$. Such a logarithm turns out, in certain circumstances, to be extendable to a homomorphism $\chi : C(p)^* \to C(k)$ for infinitely many primes p.

The key to the equidissections in Chapter 5 was the notion of a valuation ϕ. Though ϕ is not a ring homomorphism from R to R, it does behave fairly well with respect to the ring structure.

In Chapter 6, which concerns mainly tilings by similar triangles, we relied on isomorphisms between subfields of the reals or of the complex numbers. These induced certain transformations on subsets of the plane, which were also homomorphisms.

The proof of Rédei's theorem in Chapter 7 repeatedly used arguments that exploited characters. A character is yet another homomorphism, this time from a finite abelian group to the complex numbers of modulus 1. Any character can be extended to a ring homomorphism from a group ring to the field of complex numbers.

To apply algebra to a geometric problem, then, we first translated the geometric conditions into some algebraic structure. Then we worked in that structure or, with the aid of a homomorphism, in some other algebraic structure, chosen either because it is simpler or because it is richer.

These observations show that the book indeed deserves to bear the subtitle, "Homomorphisms in the service of geometry."

Appendix A
Lattices

Let $v_1, v_2, \ldots, v_n$ be n linearly independent vectors in Euclidean n-space, R^n. The set of vectors of the form $m_1v_1 + m_2v_2 + \cdots + m_nv_n$, $m_i \in Z$, is called a *lattice* of dimension n and $v_1, v_2, \ldots, v_n$ form a *basis* for the lattice. The parallelepiped spanned by $v_1, v_2, \ldots, v_n$ is a *fundamental parallelepiped* for the lattice.

For example, the vectors $v_i = (0, \ldots, 0, 1, 0, \ldots, 0), 1 \leq i \leq n$, where the 1 is in the ith place and 0's are in the other $n - 1$ places, are a basis for the lattice Z^n.

Exercise 1. When do the vectors $w_1 = (a, b)$ and $w_2 = (c, d)$ form a basis for Z^2?

Exercise 2. When is the vector $w_1 = (a, b)$ part of a basis for Z^2?

Exercise 3. Let $v_1, v_2, \ldots, v_n$ be a basis for the lattice L. Let $w_1, w_2, \ldots, w_n$ be in L. Then there are integers a_{ij} such that

$$w_i = \sum_{j=1}^{n} a_{ij}v_j.$$

For which matrices $A = (a_{ij})$ is the set $w_1, w_2, \ldots, w_n$ also a basis for L?

Exercise 4. If $v_1, v_2, \ldots, v_n$ and $w_1, w_2, \ldots, w_n$ are bases for the same lattice, show that the determinant of the matrix whose rows are $v_1, v_2, \ldots, v_n$ is equal to $+$ or $-$ the determinant of the matrix whose rows are $w_1, w_2, \ldots, w_n$.

Exercise 5. Prove that the volume of the fundamental parallelepiped associated with a basis for a given lattice L is independent of the particular basis.

Exercise 6. Let $G = Z^n$ and let H be a lattice of dimension n in G. Prove that the index of H in G is finite.

By Exercise 6, there is a finite set $g_1, g_2, \ldots, g_r$ in Z^n such that each element in Z^n is uniquely expressible in the form $h + g_i, h \in H$, $1 \leq i \leq r$. Thus translates of $\{g_1, g_2, \ldots, g_r\}$ by elements of H tile Z^n. Therefore a large cube in R^n of volume V contains approximately V/r elements of H. Let V^* be the volume of a fundamental parallelepiped of H. The large cube of volume V contains approximately V/V^* translates of the fundamental parallelepiped by translates of H. Therefore $r = V^*$. This informal argument suggests that the index of H in Z^n is equal to the volume of a fundamental parallelepiped of the lattice $H \subset Z^n$.

Exercise 7. Fill in the details to make the informal argument rigorous.

We will also need k-dimensional lattices in R^n, where $k < n$. Let $v_1, v_2, \ldots, v_k$ be k linearly independent vectors in R^n. The set of vectors of the form $m_1 v_1 + m_2 v_2 + \cdots + m_k v_k$, $m_i \in Z$, is a lattice of dimension k.

When is a subset L of R^n a lattice? Necessarily, L is a group under addition and contains no accumulation points (equivalently, every ball contains only a finite number of points of L). These conditions are also sufficient, as the following theorem shows.

Theorem 1. *Let L be a subset of R^n that is a group under addition and contains no accumulation points. Then L is a lattice.*

Proof. Since L has no accumulation points, there is a shortest nonzero vector in L. Call one such vector v_1. There are now two cases. Either L is contained in the vector space generated by v_1 or it is not.

In the first case we show that $\{v_1\}$ is a basis for L. To do this, let v be any vector in L. Then v is of the form xv_1 for some real number x. Then $x = n + f$, where $n \in Z$ and $0 \le f < 1$. Since $nv_1 \in L$, we have $fv_1 = v - nv_1$ also in L. But fv_1 is shorter than v_1, hence must be the zero vector. So $\{v_1\}$ is a basis for L.

In the second case, let Π be the plane through the origin perpendicular to v_1. Every vector $v \in R^n$ can be expressed uniquely in the form $v = xv_1 + w$, where $x \in R$ and $w \in \Pi$. Call w the projection of v on Π, and denote it $T(v)$. Let Π_1 be the plane parallel to Π through the tip of v_1, as shown in Figure 1.

Let $T(L) = \{T(v), v \in L\}$. We will show that $T(L)$ is a lattice.

Clearly $T(L)$ is a group. To show that it has no accumulation points, note that when v is written in the form $xv_1 + w$, and $x = n + f$, as defined above, then $T(v) = T(fv_1 + w)$. Thus every vector in $T(L)$ is the projection of a vector that lies between the planes Π and Π_1. Now assume that $P \in \Pi$ is an accumulation point of $T(L)$. Let B be a ball in Π around P. A cylinder with base B and

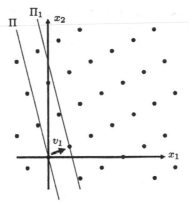

FIGURE 1

height $|v_1|$ would then contain an infinite number of points of L. This contradiction shows that $T(L)$ is a lattice.

We now may argue by induction on the dimension of the lattice. Let $w_2, \ldots, w_k$ be a basis for $T(L)$. Select $v_i \in L$ such that $T(v_i) = w_i$, $2 \leq i \leq k$. Then $v_1, v_2, \ldots, v_k$ is a basis for L. To show this, consider $v \in L$. Then $T(v) = n_2 w_2 + \cdots + n_k w_k$ for some integers $n_2, \ldots, n_k$. Hence $T(v - n_2 v_2 - \cdots - n_k v_k) = 0$. Thus $v - n_2 v_2 - \cdots - n_k v_k$ is in the space spanned by v_1. Since $v - n_2 v_2 - \cdots - n_k v_k$ is also in L, it is of the form $n_1 v_1$, for some integer n_1. This shows that $v = n_1 v_1 + n_2 v_2 + \cdots + n_k v_k$, and the inductive step is complete. $\qquad\square$

Theorem 2. *Let M be a sublattice of the lattice L. Then M is a summand of L if and only if $zl \in M$ implies $l \in M$ for every nonzero integer z and vector $l \in L$.*

Proof. Assuming that the condition is satisfied, we will show that M is a summand of L. Consider the quotient group L/M, which is a finitely generated abelian group without elements of finite order (other than the identity element). Thus L/M has a basis of the form $l_1 + M, l_2 + M, \ldots, l_r + M$. If $m_1, m_2, \ldots, m_s$ is a basis for M, then $m_1, m_2, \ldots, m_s, l_1, l_2, \ldots, l_r$ is a basis for L. Letting N be the lattice with basis $l_1, l_2, \ldots, l_r$, we have the direct sum decomposition, $L = M \oplus N$.

Now assume that M is a summand of L, $L = M \oplus N$. Assume that $l \in L$ is not in M. Then $l = m + n$, $m \in M$, $n \in N$, n not 0. If z is a nonzero integer, then zl is not 0. Thus z is not in M. This completes the proof. $\qquad\square$

Appendix B
The Character Group and Exact Sequences

The character group of a finite abelian group is used in the proof of Rédei's theorem in Chapter 7 and in the proof of Theorem 4 in Chapter 3. We define this group here and develop its properties in a sequence of exercises. Unless otherwise stated, all the groups are finite and abelian.

Let C be the group of complex numbers of magnitude 1 under multiplication and let G be a finite abelian group. A homomorphism $\chi : G \to C$ is called a *character* of G. The product of the characters χ_1 and χ_2, denoted $\chi_1\chi_2$, is defined by the equation

$$(\chi_1\chi_2)(g) = \chi_1(g)\chi_2(g),$$

for $g \in G$. With this multiplication, the set of characters form a group, called the *character group* of G, and is denoted $\overline{G}$.

Exercise 1. Prove that if $G = C(n)$, the cyclic group of order n, then $\overline{G}$ is isomorphic to G.

Exercise 2. Prove that if G is the direct product of G_1 and G_2, then $\overline{G}$ is isomorphic to $\overline{G}_1 \otimes \overline{G}_2$.

Exercise 3. Prove that any finite abelian group is isomorphic to its character group.

Any character of a subgroup of G can be extended to a character of G. This is a consequence of the following exercise.

Exercise 4.   Let H be a subgroup of the finite abelian group G and a an element of G that is not in H. Let χ be a character of H. Then χ can be extended to a character of the subgroup spanned by a and H. (Suggestion: Let n be the minimum of the set of positive integers m such that $a^m \in H$. Let z be a complex number such that $z^n = \chi(a^n)$. Define $\chi^*(a)$ to be z and $\chi^*(a^r h) = z^r \chi(h)$. Show that χ^* is well defined and extends χ.)

A sequence of abelian groups $G_0, G_1, G_2, \ldots, G_m$ with homomorphisms $\alpha_i : G_i \to G_{i-1}, 1 \leq i \leq m$,

$$G_m \xrightarrow{\alpha_m} G_{m-1} \xrightarrow{\alpha_{m-1}} \cdots \xrightarrow{\alpha_3} G_2 \xrightarrow{\alpha_2} G_1 \xrightarrow{\alpha_1} G_0$$

is *exact* if the image of α_i equals the kernel of α_{i-1}, $2 \leq i \leq m$. Consider an exact sequence of five groups where the first and last are the group of order 1:

$$\{0\} \longrightarrow A \xrightarrow{\alpha} G \xrightarrow{\beta} B \longrightarrow \{0\}. \qquad (1)$$

This sequence is exact if and only if β is a homomorphism from G onto B and α is a one-to-one isomorphism of A onto the kernel of β. Loosely put, "B is isomorphic to G/A."

If there is an exact sequence (1) it turns out that there is also an exact sequence

$$\{0\} \longrightarrow B \xrightarrow{\beta_1} G \xrightarrow{\alpha_1} A \longrightarrow \{0\}. \qquad (2)$$

This implies that for finite abelian groups a group is a homomorphic image of G if and only if it is isomorphic to a subgroup of G. (For infinite abelian groups, this is not true: $C(2)$ is a homomorphic image of Z, viewed as an additive group, but is not isomorphic to a subgroup of Z.)

Exercise 5 shows that there is the reverse exact sequence (2). For convenience, assume that A is a subgroup of G and α is the inclusion mapping.

Define $\overline{\alpha} : \overline{G} \to \overline{A}$ as follows. For $\chi \in \overline{G}$, define $\overline{\alpha}(\chi)$ to be the restriction of χ to A. Define $\overline{\beta} : \overline{B} \to \overline{G}$ as follows. For $\chi \in \overline{B}$, let $\overline{\beta}(\chi)$ be the composition $\chi\beta$.

Exercise 5. Show that the sequence

$$\{0\} \longrightarrow \overline{B} \xrightarrow{\overline{\beta}} \overline{G} \xrightarrow{\overline{\alpha}} \overline{A} \longrightarrow \{0\}$$

is exact.

Combining Exercises 3 and 5, we see that if there is an exact sequence (1), then there is an exact sequence (2).

The final exercise shows that a set of distinct characters is linearly independent.

Exercise 6. Let G be a finite abelian group and $\chi_1, \chi_2, \ldots, \chi_n$ distinct characters of G. Show that if $c_1, c_2, \ldots, c_n$ are complex numbers such that

$$\sum_{i=1}^{n} c_i \chi_i(g) = 0$$

for all $g \in G$, then $c_i = 0$, $1 \le i \le n$, as follows. If there is a nontrivial linear dependence between some of the characters, there is one with the smallest number of nonzero coefficients. Assume that

$$\sum_{i=1}^{n} c_i \chi_i(g) = 0$$

is such a relation, $c_i \ne 0$, $1 \le i \le n$. Note that n is greater than 1. Then obtain a linear dependence with fewer nonzero coefficients. (Suggestion: Pick $g_0 \in G$ such that $\chi_1(g_0) \ne \chi_2(g_0)$. Then show

$$\sum_{i=1}^{n} c_i \chi_i(g_0) \chi_i(g) = 0$$

for all $g \in G$. Using the two linear relations involving $\chi_1, \ldots, \chi_n$, produce one that has fewer nonzero coefficients.)

Appendix C
Formal Sums

Let $S = \{s_1, s_2, \ldots, s_n\}$ be a finite set and A an abelian group written additively. We will give a precise definition of the group of "formal sums," $a_1 s_1 + a_2 s_2 + \cdots + a_n s_n$, $a_i \in A$.

Let f be a function from S into A. If $f(s_i) = a_i$, $1 \leq i \leq n$, denote f by the formal sum

$$a_1 s_1 + a_2 s_2 + \cdots + a_n s_n.$$

Denote the set of the formal sums by $G(S, A)$.

Let f and g be functions from S into A. Define their sum, $f+g$, by $(f + g)(s) = f(s) + g(s)$ for each $s \in S$. There is another way to describe this addition. Suppose that $f(s_i) = a_i$ and $g(s_i) = b_i$, $1 \leq i \leq n$. Then the sum of the corresponding formal sums is defined by

$$\left(\sum a_i s_i \right) + \left(\sum b_i s_i \right) = \sum (a_i + b_i) s_i.$$

In Chapter 6 the set S consists of ordered pairs of the form PQ, where P and Q are vertices in a fixed dissection of a polygon and the abelian group A is Z. In this case $G(S, Z)$ is called the group of chains. Let H be the subgroup of G generated by the chains of the form $PQ + QP$ or $PQ + QR + RP$, where P, Q, and R are collinear and Q is between P and R.

The line integral $\int_E x \, dy$ is defined for each directed edge E and therefore $\int_C x \, dy$ is defined by linearity for each chain C. Note that if C is in H, then $\int_C x\,dy = 0$. Therefore we may view $\int x\,dy$ as being defined on the elements of the quotient group G/H.

Exercise 1.   Show that if P, Q, and R are collinear vertices, then $PQ + QR + RP$ is in H, even if Q is not between P and R.

In Chapter 7 the set S is a finite group, whose composition we denote multiplicatively and the abelian group A is Z. Note that there is a multiplication in Z. In this case $G(S, Z)$ can be made into a ring, called the group ring of the group S. Addition in $G(S, Z)$ has already been defined. Define the product of $\sum a_i s_i$ and $\sum b_i s_i$ to be $\sum c_i s_i$, where $c_i = \sum a_j b_k$, where the summation is over all pairs (j, k) such that $s_j s_k = s_i$. This definition is similar the definition of the product of two polynomials.

Appendix D
Cyclotomic Polynomials

Let n be a positive integer. The equation $x^n - 1 = 0$ has n complex roots, called nth *roots* of unity. For an nth root of unity ρ let m be the smallest of the positive integers j such that $\rho^j = 1$, that is, its order in the multiplicative group of nth roots of unity. By Lagrange's theorem, m divides n. An nth root of unity whose order is n is called a *primitive* nth root of unity. The polynomial

$$F_n(x) = \prod(x - \omega),$$

as ω runs through the primitive nth roots of unity, is called the nth *cyclotomic polynomial.* Its degree is given by the Euler phi-function, $\phi(n)$, which is the number of integers i, $1 \le i \le n$, relatively prime to n. Since each root of unity is a primitive dth root of unity for some divisor d of n, we have

$$x^n - 1 = \prod_{d|n} F_d(x). \tag{1}$$

With the aid of (1) we can show by induction that each cyclotomic polynomial has integer coefficients. (Moreover, it is irreducible over Q, as most algebra texts prove.)

Note that for a prime p,

$$F_p(x) = 1 + x + \cdots + x^{p-2} + x^{p-1},$$

since each primitive pth root of unity is a root of $(x^p - 1)/(x - 1)$.

Exercise 1. Prove that $F_n(x)$ is in $Z[x]$.

Let Ω_n be the field obtained by extending Q by the nth roots of unity, or, equivalently, by a single primitive nth root of unity. The degree of this extension, $[\Omega_n : Q]$, is $\phi(n)$, the degree of $F_n(x)$.

Lemma 1. *If $(m, n) = 1$ then the smallest field containing Ω_m and Ω_n is Ω_{mn}.*

Proof. Let $\Omega_m \Omega_n$ denote the smallest field containing Ω_m and Ω_n. Clearly $\Omega_m \Omega_n \subset \Omega_{mn}$. Now, since $(m, n) = 1$ there are positive integers a and b such that $1 = am - bn$. Let ρ be a primitive (mn)th root of unity. Then

$$\rho = (\rho^m)^a / (\rho^n)^b.$$

But ρ^m is in Ω_n and ρ^n is in Ω_m. Hence $\Omega_{mn} \subset \Omega_m \Omega_n$, and the lemma follows.   □

In Chapter 7 we use the following theorem, which is an immediate consequence of Lemma 1.

Theorem 1. *If $(m, n) = 1$, then $F_m(x)$ is irreducible in the ring $\Omega_n[x]$.*

Exercise 2. Prove the theorem.

Exercise 3. Show that if p is a prime and n is a positive integer, then $F_{p^n}(x) = F_p(x^{p^{n-1}})$. (Hint: If ρ is a primitive (p^n)th root of unity, show that $\rho^{p^{n-1}}$ is a primitive pth root of unity.)

Bibliography for Preface

1. F. W. Barnes, Algebraic theory of brick packing I, *Discrete Math.* **42** (1982), 7–26, and II, *ibid*, 129–144.

2. R. Berger, *The undecidability of the domino problem*, Memoirs Amer. Math. Soc. No. **66** (1966), 72 pp.

3. V. G. Boltanskii, *Hilbert's third problem*, Scripta Series in Mathematics, Wiley, New York, 1978. (Review in *Bull. AMS*, **1** (1979), 646–650.)

4. ——, *Equivalent and equidecomposable figures*, D. C. Heath, Lexington, 1963.

5. J. H. Conway and J. C. Lagarias, Tiling with polyominoes and combinatorial group theory, *J. Comb. Theory Series A* **53** (1990), 183–208.

6. N. G. de Bruijn, Algebraic theory of Penrose's non-periodic tilings of the plane I, *Proc. Akad. van Wetenschappen, Proc. Ser. A* **84** (1981), 39–52, and II, *ibid.*, 53–66.

7. F. M. Dekking, Replicating superfigures and endomorphisms of free groups, *J. Comb. Theory Series A* **32** (1982), 315–320.

8. D. Gale, More on squaring squares and rectangles, *Mathematical Intelligencer* **15**, number 4 (Fall, 1993), 60–61.

9. S. Golomb, *Polyominoes*, Princeton University Press, 1994.

10. B. Grünbaum and G. C. Shephard, Tiling with congruent tiles, *Bull. Amer. Math. Soc.* **3** (1980), 951–973.

11. ——, *Tiling and Patterns*, Freeman, New York, 1987.

12. R. Kenyon, Tiling with squares and square-tileable surfaces, Ecole Normale Supérieure de Lyon, **119** (1993), 1–26.

13. D. A. Klarner and F. Göbel, Packing boxes with congruent figures, *Indag. Math.* **31** (1969), 465–472.

14. M. Laczkovich and G. Szekeres, Tilings of the square with similar rectangles, Conference in honor of László Fejes-Tóth's 80th Birthday, to appear in *Discrete and Combinatorial Geometry.*

15. D. G. Mead and S. Stein, More on rectangles tiled by rectangles, *Amer. Math. Monthly* **100** (1993), 641–643.

16. H. Meschkowski, *Unsolved and unsolvable problems in geometry*, Ungar, New York, 1966.

17. R. Penrose, Pentaplexity, a class of non-periodic tilings of the plane, *Math. Intelligencer* **2** (1979), 32–37.

18. R. M. Robinson, Undecidability and nonperiodicity of tilings of the plane, *Inventiones Mat.* **12** (1971), 177–209.

19. D. Schattschneider, Will it tile? Try the Conway criterion, *Math. Magazine* **53** (1980), 224–233.

20. S. Stein, The notched cube tiles R^n, *Discrete Math.* **80** (1990), 335–337.

21. S. Szabó, Cube tilings as contributions of algebra to geometry, *Contributions to Algebra and Geometry* **34** (1993), 63–75.

22. W. P. Thurston, Conway's tiling groups, *Amer. Math. Monthly* **97** (1990), 757–773.

23. S. Wagon, Fourteen proofs of a result about tiling a rectangle, *Amer. Math. Monthly* **94** (1987), 601–617.

Name Index

Subject Index

Symbol Index